D. Kottonau, E. Shabagin, W. de Sousa, J. Geisbüsch,
M. Noe, H. Stagge, S. Fechner, H. Woiton, T. Küsters

Evaluation of the Use of Superconducting 380 kV Cable

Karlsruher Schriftenreihe zur Supraleitung Band 030

HERAUSGEBER

Prof. Dr.-Ing. M. Noe
Prof. Dr. rer. nat. M. Siegel

Eine Übersicht aller bisher in dieser Schriftenreihe erschienenen Bände finden Sie am Ende des Buches.

1 Motivation and task

The authors would like to thank the Federal Ministry of Education and Research (BMBF) for the funding provided within the Copernicus project ENSURE "Neue EnergieNetzStruktURen für die Energiewende" (New Energy Grid Structures for the Energy Turnaround) (funding code 03SFK1L0). Responsibility for the content of this study lies with the authors.

1.1 Motivation

The ongoing and planned conversion of the electrical energy supply to a new generation structure requires wiring and substations to be upgraded at many points in the German extra-high voltage grid. In addition, some completely new wiring will be planned and installed.

Many of the measures are running in the conventional 380 kV three-phase grid. In addition, long DC lines will support electrical power transmission; these connections will be implemented as point-to-point connections.

In the three-phase network, new lines will continue to be constructed predominantly as overhead lines. Underground cable route sections can only be realised in legally prescribed pilot lines under certain conditions. The following section describes exactly which conditions must be fulfilled for this.

Passively cooled, plastic-insulated cables are predominantly installed on the underground line sections. Alternative technologies, such as actively cooled plastic cables, gas-insulated cables and superconducting cables, are currently still in the research and development stage or are being tested on selected lines and are therefore not available for general use in the extra-high voltage grid.

This study describes a layout of superconducting cables for use in the 380 kV three-phase AC grid and explains general aspects of the use of such cables in the extra-high voltage grid. It compares superconducting technology with other line technologies under many different criteria.

1.2 Possible operation sites

The use of underground transmission technologies in the German alternating current high-voltage grid is only possible under certain conditions in pilot projects directly referred to in the Federal Requirements Plan Act or in the Energy Line Expansion Act. It is legally stipulated that the proximity to housing estates or expected nature conservation conflicts, for example impairments of a "Natura 2000" area or prohibitions under species protection law, require the examination of an underground laying in some sections of the route.

In addition, the law stipulates that an underground cable line must only be constructed on "technically and economically efficient sub-sections of the route". This means that the such sub-sections should not exceed or fall short of certain lengths. For very short sections, the transition system required on both sides of the underground section causes high costs and a large amount of land consumption; for very long sections, the total costs are very high compared to the overhead line variant, as an overhead line costs considerably less per km. normally, underground cable sections should therefore be planned with a length of 3-10 km.

As part of the planning procedure for the pilot projects, the responsible transmission system operator must check whether the statutory conditions for a partly underground cable are met for line sections and whether the advantages over an overhead line, for example for the protection of the residential environment, are large enough to justify the design as an underground cable line. The final decision on the assignment of a partly underground cable is taken by the competent authority in the planning approval procedure.

The lists below include the pilot projects name in the Federal Requirements Plan Act and in the Energy Line Expansion Act, in which the use of underground cable is to be reviewed.

Federal Requirements Plan Act:

- No. 6: Extra-high voltage line Conneforde - Cloppenburg East - Merzen
- No. 7: Extra-high voltage line Stade - Sottrum - Wechold - Landesbergen
- No. 31: Extra-high voltage line Wilhelmshaven - Conneforde
- No. 34: Extra-high voltage line Emden East - Conneforde
- No. 42: Extra-high voltage line district Segeberg - Lübeck - Siems - Göhl

Energy Line Expansion Act:

- Section Ganderkesee – St. Hülfe of the line Ganderkesee – Wehrendorf
- Line Diele – Niederrhein
- Wahle – Mecklar line
- Section Altenfeld – Redwitz of the line Lauchstädt – Redwitz
- Rhine crossing in section Wesel - Utfort of the line Lower Rhine - Utfort - Osterath
- Line Wehrendorf – Gütersloh

1.3 Investigated installation

Figure 1.1: Overview of a partly underground cable in conventional and superconducting design shows the overview of a partly underground cable in conventional and superconducting design.

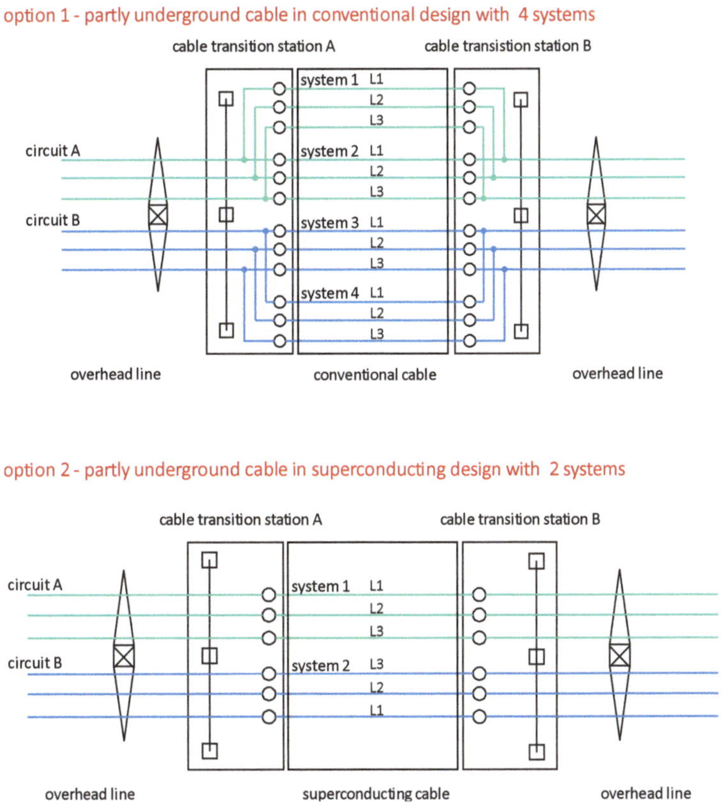

Figure 1.1: Overview of a partly underground cable in conventional and superconducting design

Terminations are required at both ends of the cable for the transition between overhead line and cable section. Additional equipment and components are often required. All this equipment is built into a cable transition system.

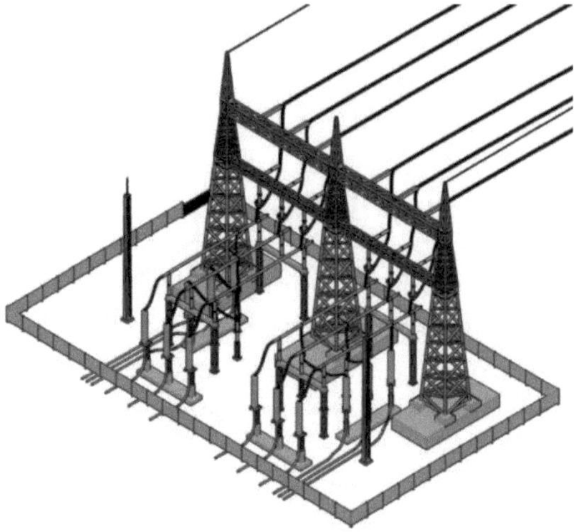

Figure 1.2: Representation of a cable transition system.

In this study, a superconducting cable system is planned as example for an intermediate cabling section approved by TenneT and currently under construction. The section is located on the line Diele - Niederrhein near the city of Haren (Ems). Over a length of approx. 3.2 km the overhead line will be replaced by a conventional underground cable system. The cable section is shown in the appendix (Figure A.17).

1.4 Network requirements for the cable system

The maximum current carrying capacity and thus the maximum electrical power to be transmitted serve as the central design parameters for a line. In the German extra-high voltage grid, new lines are designed according to the planning premises of the German transmission system operator from 2017 for 3,600 A per circuit. The 3,600 A are particularly required for the emergency operation of a line; the usual operating current is usually well below this value.

The electrical behaviour of underground wiring in the network should ideally be as close as possible to the known behaviour of the overhead line, because the installations, industrial plants, controls and operating concepts used today are designed for overhead lines. This concerns the impedance of the line, which should not have a high capacitance, as well as the natural performance and the fault behaviour or the required protection settings.

Another important aspect is the availability of a line, which depends largely on the technology used. In recent years, the failure statistics for the electrical power supply in Germany show a constant availability of 99% with average downtimes of around 15 minutes per year. In this respect, underground cable line technologies must be measured against the extremely high availability of the overhead line. A worse availability of line sections should not lead to the electrical energy supply in Germany becoming noticeably less reliable or to further wiring becoming necessary as additional redundancy.

1.5 Line types in comparison

This study focuses on a comparison of the technical parameters. It focuses on the properties of superconducting cable and classifies it in comparison with other transmission technologies.

The technical properties of the various transmission technologies are already listed in several studies. Therefore, this study gives only a brief overview. Further information can be found in the respective studies; especially the DENA study contains an excellent overview [DEN14]. A comparison of the socio-ecological properties, such as the emission of fields, is omitted.

1.5.1 Overhead line

The overhead line is the predominant transmission technology in today's German extra-high voltage grid with over 99% of the total line length. Its main disadvantage is that it is installed above ground and can be seen over long distances due to the required air gaps and the resulting mast heights.

The above-ground installation and the use of air as an insulating material also offer important advantages: On the one hand the electrical capacity of overhead lines is relatively low, on the other hand troubleshooting and repair can be carried out quickly after a defect so that the line can go back into operation quickly.

Overhead line damage caused by the weather and other above-ground events, such as mast breakage during storms or conductor breakage during heavy ice loads, occur only very rarely in Germany; lightning strikes during thunderstorms can usually be passed through by automatic reclosing (AWE; also known as short interruption).

1.5.2 Conventional cable

To transmit the required current intensity, two systems of plastic insulated cables must be used per circuit. Both systems will be installed in shallow laying. Figure 1.3 shows the dimensions of a cable szstem as example.

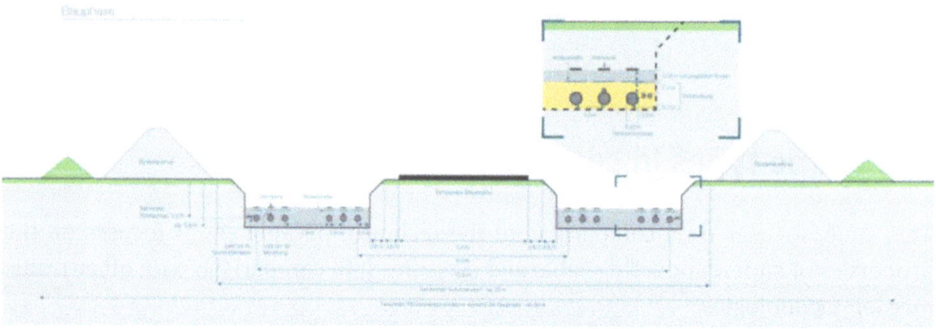

Figure 1.3: Overview of the cable trench profile for two three-phase current systems on the 380 kV voltage level when using copper cables with a conductor cross-section of 2500 mm^2..

1.5.2.1 Design of the underground cable

The width of the cable system depends on the thermal design. The waste heat generated by the losses during operation must be dissipated via the ground so that the cable system does not overheat.

This study only provides a loss calculation as a comparison; further details on the design of plastic insulated cable systems can be found in the literature.

1.5.2.2 Conductor losses

According to equation 1.1, the conductor loss layer $P'_{\mathrm{V,EK,g}}$ of a subconductor depends on the resistance per unit length R'_{EK}, the number of systems n_{SYS}, the conductor current I_{PH}.

$$P'_{\mathrm{V,EK,g}} = n_{\mathrm{SYS}} \cdot \underbrace{3 \cdot R'_{\mathrm{EK}} \cdot I_{\mathrm{PH}}^2}_{P'_{\mathrm{V,EK,SYS}}} \tag{1.1}$$

To simplify matters, a value for ohmic resistance is only assumed that is temperature-independent and takes into account the skin and proximity effects. The cables with a cross-section of 2500 mm^2 copper have a resistance per unit length R'_{EK} of 0.0109 Ω/km; this results in the power losses indicated in Table 1.1.

Table 1.1: Overview of the effective losses at different load factors for four systems

Cable type	Symbol	Power loss $0{,}1 \cdot I_{\mathrm{N}}$ in W/m	Power loss $0{,}5 \cdot I_{\mathrm{N}}$ in W/m	Power loss $1 \cdot I_{\mathrm{N}}$ in W/m
2XSFL2Y-2500 mm^2	$P'_{\mathrm{V,EK,g}}$	4.2	106.0	423.8
	$P'_{\mathrm{V,EK,SYS}}$	1.1	26.5	106.0

1.5.2.3 Dielectric losses

The dielectric losses $P_{\mathrm{V,0,EK}}$ of a cable are calculated in equation 1.2.

$$P_{\mathrm{V,0,EK}} = U_{\mathrm{LE}}^2 \cdot \omega \cdot C'_{\mathrm{EK}} \cdot tan\delta \tag{1.2}$$

$$C'_{\mathrm{EK}} = \frac{\epsilon_{\mathrm{ISO}}}{18 \cdot ln\left(\frac{d_{\mathrm{ISO}}}{d_{\mathrm{in,LS}}}\right)} \tag{1.3}$$

U_{LE} Conductor-earth voltage $U_{\mathrm{LE}} = U/\sqrt{3}$ (V)
ω Angular frequency $\omega = 2\pi f$ (1/s)
C'_{EK} Capacitance per unit length (F/m)
$tan\delta$ Factor of loss of insulation according to IEC 60287 table 3ϵ_{ISO}Relative permittivity of insulation according to IEC 60287 table 3
d_{ISO} Diameter over insulation without external conducting layer (mm)
$d_{\mathrm{in,LS}}$ Diameter over inner conducting layer (mm)

The resulting capacitances per unit length and dielectric losses are shown in Table 1.2. The dielectric losses are load-independent, so they are incurred over the entire operating time.

Table 1.2: Overview of capacitances per unit length and dielectric losses for four systems

Cable type	Symbol	Capacitance per unit length F/m	Dielectric losses W/m
2XSFL2Y-2500 mm^2	$P'_{\mathrm{V,0,EK,g}}$	$2{,}33 \cdot 10^{-10}$	42.4
	$P'_{\mathrm{V,0,EK,SYS}}$		10.6

1.5.2.4 Overview of the total losses in underground cables

The total losses, represented in Table 1.3, are made up of the conductor losses and the dielectric losses.

Table 1.3: Overview of losses for four systems with a rated current of 1.8 kA per system and a route length of 3200 m for 2500 mm² cross-section

Loss component	Symbol	Power loss	Power loss	Power loss
		$0{,}1 \cdot I_N$	$0{,}5 \cdot I_N$	$1 \cdot I_N$
Resistive losses	$P_{V,EK,g}$	14 kW	339 kW	1356 kW
Dielectric losses	$P_{V,0,EK,g}$	136 kW	136 kW	136 kW
Total losses	$\boldsymbol{P_{V,g}}$	149 kW	475 kW	1492 kW

1.5.3 Gas insulated lines

Gas insulated lines (GIL) are a possible alternative to VPE insulated cables. They promise higher transmission performance per system and can therefore be made narrower.

With a gas-insulated line, an insulating gas is installed in a jacket pipe (normally SF6/N2 with 10-20 % SF6 and a filling pressure of 7 bar). The aluminium inner conductor is fixed in the middle of the jacket pipe by means of epoxy resin support insulators.

The current-dependent losses amount to about one third of the losses of overhead line. Dielectric losses are negligible with gas-insulated wiring. Gas insulated lines have an electrical capacity three to four times lower than VPE cable. Compared to double cable systems with the same power, the capacity of the GIL is only about one seventh of that of the XLPE cable system; this significantly reduces the need for compensation for reactive power. In principle, gas-insulated lines have similar operating characteristics to overhead cables; with regard to protection and control systems, there is no difference to the use of overhead cables. The operation with automatic reclosing is also possible.

Underground cable gas insulated wiring will be laid normally at a depth of about 1.5 m. The route width for two systems in operation is about 8 m for two GIL systems.

In Germany, a tunnel-laid GIL of the first generation of 700 m length has been in operation since 1976. An underground GIL line - two systems with 1,800 MVA each - with a length of 900 m was put into operation at Frankfurt Airport in 2010. In 2014, Munich municipal utility put a tunnel-laid GIL of approx. 450 m length into operation.

1.5.4 Superconducting cable

Superconducting cables promise the highest transmission performance due to their high current intensity. important properties of a superconducting cable are the low electrical losses at high load, the freedom from field due to the magnetic shielding and the low route width requirement due to the high current intensity per single cable and the active cooling. A detailed layout of a superconducting cable concept is described in the following chapter.

1.6 Scope of examinations

The scope of the investigations in this study is derived from the task of carrying out a technical and economic evaluation of a superconducting 380 kV cable at a real existing location.

For this purpose, section 1 first deals with the location and the network requirements derived from it, then with possible line types.

Section 2 describes the basic setting up of the superconducting high voltage cable based on the function and mode of operation of all individual dynamics. In addition to the cable components, this includes the cooling, the cable ends and the connection of two cables.

Based on this, section 3 contains the detailed layout including the individual design steps and the parameters and formulas used.

The comparison of the electrical line capacity and the resulting significant property is summarized in section 4.

Section 5 examines in detail the cost-effectiveness of using a superconducting cable and compares it with that of a conventional solution. This includes a detailed description of the calculation of the operating costs, based on the calculation of the annual loss costs at different load levels, as well as an overview of the investment cost shares.

The failure behaviour of superconducting high voltage cables is compared in section 6 with that of conventional cables. In addition to fault scenarios and fault frequency, repair and spare parts concepts are also considered.

Section 7 summarises the results and evaluates them.

2 Set-up and status of the development of superconducting cable systems for 380 kV cable systems

2.1 Overall system

Superconducting cable systems can have a higher current carrying capacity than conventional cable systems because the superconducting materials have high current densities of approx. 1,000-10,000 A/mm^2. In contrast, copper- and aluminium-based cable systems have current density of 1-5 A/mm^2. Superconducting materials are mainly bismuth strontium calcium copper oxide (BSCCO) and yttrium-barium copper oxide (YBCO). These materials are particularly characterized by their high critical currents and the possibility of simplified cooling with liquid nitrogen[1].

The first generation (1G) of industrially manufactured conductors with BSCCO as superconducting material requires a relatively high proportion of silver as matrix material. This silver determines the price of the conductor and thus prevents an industrially economical production. Conductor with YBCO as the superconducting material, on the other hand, requires very little silver in a very thin layer for stability. They are called superconductors of the second generation (2G).

The scaling of the processes in industrial production is successfully promoted. The conductors currently available on the market are up to 800 m long. The approximately ten manufacturers of conductors made of high-temperature superconductor materials worldwide offer their products in the form of thin strips with widths between 2 mm and 12 mm [CIG15]. These conductors, referred to as tapes or SC tapes, can be used for a wide range of electrical equipment without great technical effort; this also includes use in cable systems. Whereas only so-called single-conductor cables can be used for high and extra-high voltage cables, three-wire medium-high voltage cables or three-phase concentric cables are installed.

Generally, superconductor cables are distinguished between cold and warm dielectric. If the electrical cable insulation - the dielectric - is in the cooling circuit and is

[1] These materials are also referred to as high-temperature superconductors (HTS) to distinguish them from helium-cooled superconductor materials.

therefore also cooled, it is a cold dielectric. If the electrical cable insulation is outside the cooling circuit, it is a warm dielectric.

If the electrical isolation of a single-conductor cable is cooled, a magnetic field shielding can be added to the cable design, which is referred to as HTS shielding or HTS shield. This shielding layer can suppress an external magnetic field complete with the appropriate layout. With three-phase concentric cables, the magnetic fields cancel each other out when the load is symmetrically distributed. A further shielding layer is therefore not necessary in this case.

Insulated cables with warm dielectric cannot be shielded in the same way. The advantage of cables with warm dielectric is that they can be equipped with a conventional cable insulation. In addition, the dielectric losses of the cable do not get into the cooling medium and, unlike the cable with cold dielectric, do not have to be compensated by the chiller. The following FigureFigure 2.1 shows the three cable types with their areas of application.

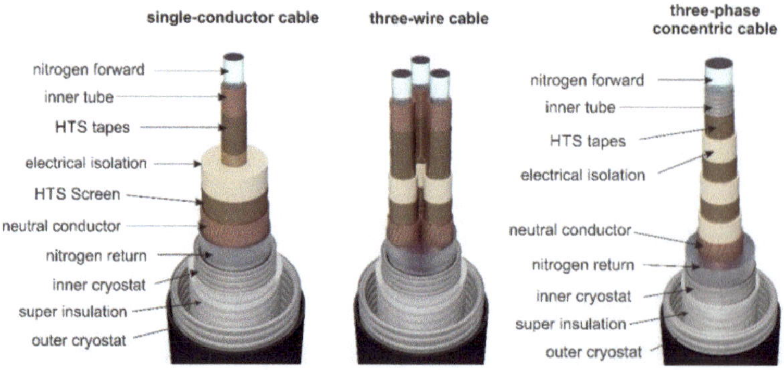

	Single-conductor cable	Three-wire cable	Three-phase concentric cable
Voltage level	High and extra-high voltage	Medium-high voltage	Medium-high voltage
Dielectric	Warm and cold	cold	cold
Superconducting requirement for three phases	similar to 3-wire cable (3 x HTS conductor+ 3 x HTS shield)	similar single-conductor cable	least (3 x HTS conductor)
Cable cryostat for three phases	3 cryostats (3 outer tubes and 3 inner tubes)	1 cryostat (1 outer tube, 1 inner tube)	1 cryostat (1 outer tube, 1 inner tube)

Figure 2.1: Schematic representation of the structure and application areas of single-conductor cables, three-conductor cables and three-phase concentric cables.

Since the study examines the 380 kV voltage level, it deals exclusively with the general setup and the geometric dimensions of a single-conductor cable.

2.2 Cable build-up

This section describes the basic layer structure of a single-conductor cable. The superconducting cable consists of two main functional components: the cable core and the line cryostat. The cable core performs all electrical functions, such as power transport, electrical insulation, magnetic display and neutral conductor. The line cryostat is used for thermal insulation. The actual coolant flow depends on the cooling concept - one-sided or double-sided coolant supply - and the cooling options. This is described in more details in section 3.1.2.1.

2.2.1 Former with regulating layer

The inner tube, also referred to as former, has three tasks: First, it acts as mechanical support for the individual layers applied to it, such as conductor layer, electrical insulation layer, HTS shield layer and neutral conductor. Second, in this study it is used to transport liquid nitrogen. Third, the former carries part of the short-circuit current in the event of a short-circuit fault and thus ensures thermal stability. A combination of stainless-steel strips, metal foil and padding (polycarbonate) is applied to the corrugated inner tube to protect against deformation and damage to the SC tapes. This study examines two former concepts, described in detail in section 3.1.

2.2.2 HTS layer

The reviewed second-generation SC tapes have a layer structure consisting of a metal foil of 50-100 μm thickness as substrate, a buffer layer of various ceramic materials, a superconductor layer of approx. 1-2 μm thickness for example of yttrium-barium copper oxide, as well as a thin silver layer and copper stabilization. Width and current density vary depending on manufacturer and are 2-12 mm or 250-600 A/cm. The SC tapes are wrapped directly next to each other, if possible, without any gap, on the former with compensation layer (Figure 3.13). It is possible to use several layers of the superconducting SC tape to increase the current carrying capacity.

Figure 2.2: Representation of the layer structure of a superconducting SC tape [FUJ17].

2.2.3 Electrical insulation with semiconductor layer

The two dielectrics Polypropylene Laminated Paper (PPLP™) or high-density polyethylene (Tyvek™) can be used as electrical insulation material in liquid nitrogen. This study considers exclusively Tyvek™ for the electrical insulation, as the dielectric losses of PPLP™ are greater [KSN17].

2.2.4 HTS shield with neutral conductor layer

The protective shield is mounted on the dielectric, which shield also consists of a superconductor and is used as a return conductor. The current in the shield fully compensates the magnetic field outside the cable. As a result, no mechanical forces occur between the phases. To completely compensate the magnetic field around the conductor, the superconducting screen is used as a return conductor as shown in the figure Figure 2.3.

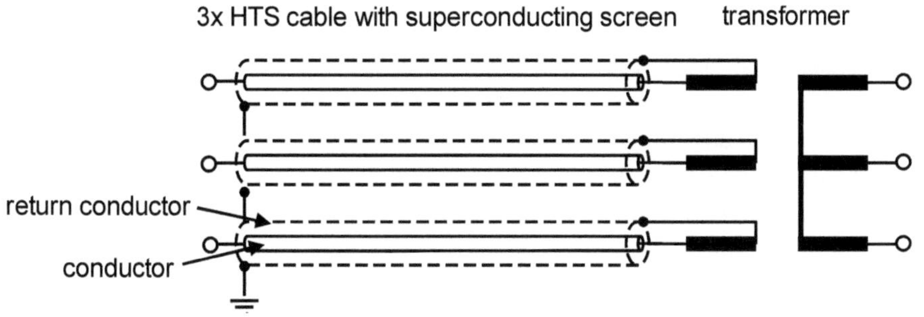

Figure 2.3: Circuit diagram for complete compensation of the magnetic field in a single-conductor cable with cold dielectric.

The neutral conductor made of copper is wrapped around the HTS layer, separated by a layer of padding. It is used to conduct the charging, leakage and fault currents.

2.2.5 Thermal insulation - cable cryostat with super insulation

The line cryostat comprises the inner and outer cryogenic tube, the super insulation with evacuated space and a polyethylene jacket on the outer cryogenic tube. The super insulation is wrapped on the outside of the inner cryostat tube and consists of 10 to 30 layers of polyester foil, which is vaporized with aluminium. This tube construction can be in flexible or rigid design. Figure 2.4 shows the schematic cross-sectional views of the flexible and rigid line cryostats.

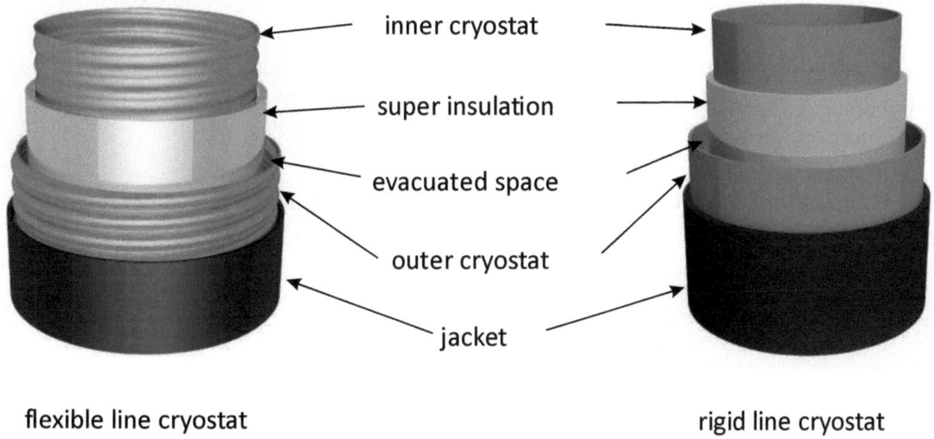

Figure 2.4: Representation of the structure of a flexible and a rigid line cryostat with super insulation.

For thermal and manufacturing reasons, the distance between the inner and outer cryostat tube must not be too small. Two spiral spacers on opposite sides ensure a coaxial position of the two cryostat tubes. Compared to flexible line cryostats, rigid cryostats have the advantage of lower thermal losses [CRY12], [CRY15]. Two inner tube variants - corrugated tube concept and hollow core conductor concept - can be combined with the two cable cryostat variants - flexible and rigid. This study covers all four combinations.

2.2.6 Termination

The terminations are used to connect an overhead line or a busbar to the superconducting cable and are attached at both ends. Superconducting single-conductor cables use a separate termination for each cable. A separate redundant cooling of the terminations is necessary to thermally and hydraulically decouple the cooling circuits. The tasks of a termination for superconducting cable are:

- electrical connection of the superconducting cable with the network
- absorption of the temperature gradient between room temperature and the temperature of the liquid nitrogen
- Electrical isolation
- compensation of contraction during cooling and heating

The design of a termination is shown in the following figure Figure 2.5. The vertical part contains the connection to the conventional network and the high-voltage insulation. The high voltage insulation contains the power supply, which absorbs the temperature gradient. The horizontal part contains the connection of the conventional conductor with the superconductor as well as the connection to the cooling system and the feeding of the liquid nitrogen into the cable. The electrical isolation for the nitrogen, which is at high voltage potential in the cable, is located between the cable and the cooling system.

The terminations for the investigated voltage level of 380 kV have not yet been realised in any project.

2.2.7 Sleeve

Sleeves are used to connect parts of a superconducting cable system. These are also made of superconducting SC tapes.

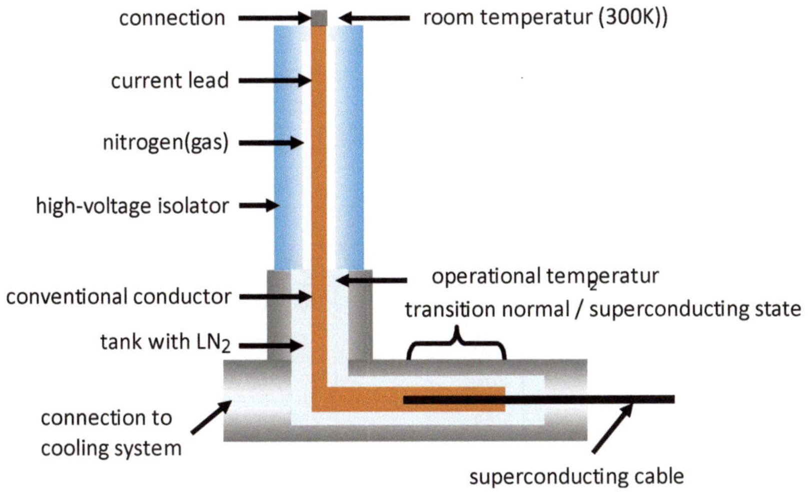

Figure 2.5: Schematic representation of a termination for superconducting high-voltage cables.

2.2.8 Further development

As shown in Table 2.1, many superconducting cables have already been successfully tested in the network. The cable with the longest continuous operating time without interruption went into operation in Essen in April 2018. It has a length of 1 km, a voltage of 10 kV and an output of 40 MVA.

Table 2.1: Overview of selected superconducting cables used in the in network

Place, manufacturer	Year	Length	Superconductor	Data
Essen, Nexans	2014	1000 m	BSCCO	10 kV, 40 MVA
Yokohama, Sumitomo	2013	250 m	BSCCO	66 kV, 200 MVA
New York, Ultera	2013	300 m	REBCO	13.8 kV, 96 MVA
Long Island, Nexans	2012	600 m	BSCCO/REBCO	138 kV, 574 MVA
Icheon, LS Cable	2011	500 m	REBCO	22.9 kV, 50 MVA

SC tapes based on BSCCO or REBCO connections are used as high-temperature superconductors. BSCCO SC tape in good quality are only manufactured by one manufacturer, Sumitomo. In recent years, however, more and more companies have been manufacturing REBCO SC tapes and offering them commercially. Table 2.2 provides an overview of this. The manufacturers differ above all in the manufacturing processes and the material compositions used.

Table 2.2: Overview of manufacturers of REBCO SC tapes

Name	Country	Ic (77K, sf)/cm	Single lengths
AMSC	USA	400-500 A	some 100 m
Bruker	Germany	~ 400 A	up to 300 m
d-nano	Germany	> 250 A	> 100 m
Fujikura	Japan	400-500 A	up to 1000 m
Shanghai Creative Superconductors	China	300-500 A	some 100 m
Shanghai Superconductors	China	300-600 A	up to 300 m
STI	USA	250-500 A	-
Sunam	Korea	300-500 A	up to 1000 m
Superpower/Furukawa	USA	200-300 A	some 100 m
Superox	Russia/Japan	up to 400 A	up to 300 m
Theva	Germany	500-600 A	some 100 m

Commercially available current densities are already many times higher than 500 A per cm conductor width at a temperature of 77 K and in the intrinsic magnetic field. In order to compare the costs for the conductors of different manufacturers, they are normally given in € per kA current carrying capacity and based on a unit length of 1 m. The costs at a temperature of 77 K vary strongly between 50 k€A m and over 200 k€A m depending on the specification and manufacturer. For comparison: Copper with a current density of 1 A/mm^2 has cost between 20 k€A m and 60 k€A m in recent years. Most manufacturers are still committed to production capacities of less than 100 km per year. A further upscaling of production will enable a significant reduction in costs, as the share of material costs is low. For the future, a further increase in the current carrying capacities is to be expected, since more than 1500 A/cm of critical current density have already been realised on a laboratory scale. Single piece lengths of the conductor are no longer available in 1000 m. However, several manufacturers have a process for connecting two low-resistance conductors so that almost any length can be produced.

In addition to further developments and ongoing cost reductions in the manufacture of SC tapes, improvements to the line cryostats have great potential for advancing superconducting transmission systems. As explained in the following sections, cryostat transmission losses account for a large proportion of the total losses of the system.

3 Conceptual design of superconducting cable systems for use in the 380 kV grid

The objective of this section is to set out a cable design that meets the network requirements. In addition, the design limits are determined by the variation of construction parameters and cooling concepts. The design parameters consider three nominal diameters of the former: DN 32, DN 40 and DN 50. Furthermore, two former concepts are examined for their short-circuit strength, taking into account three nominal values. The cooling concepts differentiate between one-sided and two-sided coolant supply. With the two cooling concepts to be investigated, this results in a total of twelve cable designs. All cable designs are tested for hydraulic, electrical and thermal properties; the design limits are considered in dependence on the mass flow of the coolant.

Section 1.4 sets out the network requirements for the cable system. These include a current carrying capacity of 3.6 kA and a route length of 3.2 km. Other operating parameters are the nominal voltage of 380 kV, the reference voltage of 420 kV and the sustained short-circuit current of 63 kA for a short-circuit duration of 300 ms. Table 3.1 provides overview of the operating parameters.

Table 3.1: Overview of the specified operating parameters for the partial underground cabling

Label	Symbol	Values
Rated voltage	U_m	420 kV
Nominal voltage	U_N	380 kV
Rated current	I_r	3.6 kA
Steady short-circuit current	I''_{kQ}	63 kA
Fault explanation times	t_{KS}	300 ms

It should be noted that in contrast to the conventional cable system in section 1.4, the rated current is selected differently. Due to their thermal capacity, conventional cables can carry a relatively high overload for several hours. Such an overload is not possible with superconducting cables, therefore the rated current must be precisely defined.

Section 3.2 investigates the cable properties with regard to resistive, inductive and capacitive equivalent parameters. In addition, a detailed loss calculation of a cable design is carried out. The impact of the additional thermal stabilisation of the hollow core conductor concept is analysed in section 3.3. A summary of the final cable design can be found in section 3.4. Based on the final cable design, section 3.5 develops an exemplary cooling system and describes central processes. The different ways of covering the own cooling requirement are explained in the section 3.6.

3.1 Cable build-up

The cable build-up section deals with the creation of the cable geometry, the calculation of the temperature and the pressure curves as well as the determination of the superconductor requirement.

3.1.1 Geometry

For a superconducting cable, the functions of the individual layers and their materials are described in detail in section 3.2. This section defines the concrete layer structure for further calculations. For this purpose, the electrical insulation thickness is designed on the basis of maximum electrical field strengths. The thickness of the electrical insulation is measured iteratively so that a design field strength of 12.5 kV/mm is achieved at the inner conductive layer and 6.5 kV/mm at the outer conductive layer. This design field strength was adopted from conventional 380 kV cable systems [Sta17]. The calculation 3. 1 is used to calculate the electric field strength.

$$\vec{E}(r) = \frac{U_0}{r \cdot ln\left(\frac{r_a}{r_i}\right)} \cdot \vec{e}_r \tag{3.1}$$

$\vec{E}(r)$	Electrical field strength	(V/m)
U_0	Operating voltage 420 kV	(V)
r	Radius at the required point ($r = r_i$ or $r = r_a$)	(m)
r_a	Outer radius (e.g. outer conductive layer)	(m)
r_i	Inner radius (e.g. inner conductive layer)	(m)

To calculate the electric field strength, the thickness and/or outer diameter r_a of the electric insulation and the inner diameter r_i of the insulation are required. This corresponds to the outer diameter of the former, without considering the semiconductor layer. The outer diameter of the former in turn varies depending on the nominal diameter of DN 32, DN 40 and DN 50 as well as on the choice of former concept - hollow core conductor concept or corrugated pipe concept. In Figure 3.1 und Figure 3.2 the outer diameters of the former variants are listed as a function of the nominal diameters. Section 3.1.2.3 explains further details on the former concepts. The calculated electric field strengths are listed as a function of the nominal diameter of the inner tube of the hollow core conductor concept in Table 3.2.

Table 3.2: Overview of the electric field strengths as a function of the nominal diameter of the inner tube of the hollow core conductor concept and the electric insulation thickness.

Nominal diameter of inner tube	DN 32	DN 40	DN 50
Thickness of the electrical isolation (Tyvek™)	25.85 mm	24.15 mm	24.25 mm
Electrical field strength-external conducting layer	5.6 kV/mm	6.3 kV/mm	6.5 kV/mm
Electrical field strength-internal conducting layer	12.5 kV/mm	12.5 kV/mm	11.9 kV/mm

After the thickness of the electrical insulation has been dimensioned, the other layers of the cable core are constructed layer by layer. The description of the former and the electrical insulation is followed by the HTS shield layer and the neutral conductor layer in the structure of the cable core. For the HTS shield layer described in section 3.1.4, a layer thickness of 0.2 mm is assumed; which corresponds roughly to the thickness of a layer of superconducting SC tape [FUJ17]. A thickness of 2 mm is assumed for the neutral conductor layer; this results in a minimum cross-sectional area of approx. 600 mm^2. Section 3.3 shows that this cross-sectional area in combination with the cooling withstands the maximum short-circuit current. A detailed dimensioning and optimization of the neutral conductor layer are not part of this study.

For a complete description of the cable design, the description of the cable core is followed by the design of the line cryostat. It is hereby assumed that the line cryostat has the structure described in section 2.2.5. The corrugated tube dimensions from Table 3.5 are used accordingly. The distance between the inside and the outer cryostat tube results from the discrete gradations of the standardized nominal value (DN 80, DN 100, DN 125 etc.). The minimum distance between the cryotubes is assumed to be 10 mm [CRY12]. Like a conventional underground cable, the superconducting cable is also surrounded by a PE sheath layer. A layer thickness of 5 mm is assumed for this [Süd 16].

In Figure 3.1 the quarter cut and the diameters of a conceptual single-core cable are shown in the hollow core conductor concept. This single-conductor cable is investigated with three possible nominal values in order to visualize the variables influencing temperature and variation of pressure.

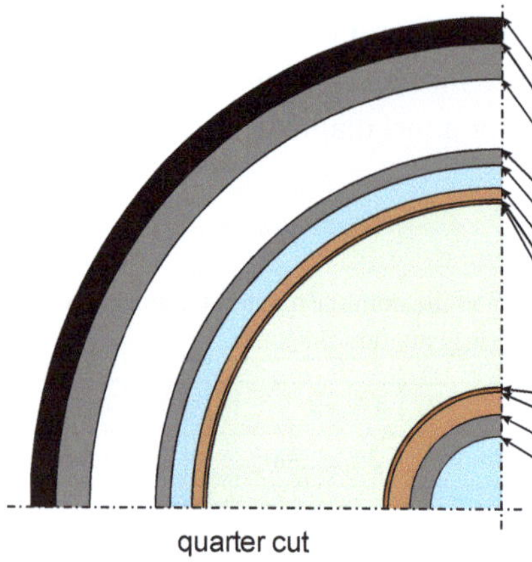

description	symbol	diameter in mm former nominal - DN 32	former nominal - DN 40	former nominal - DN 50
jacket (OD)	$d_{M,a}$	172,0	172,0	172,0
jacket (ID)	$d_{M,i}$	162,0	162,0	162,0
outer cryostat tube (ID)	$d_{AK,i}$	150,0	150,0	150,0
inner cryostat tube (OD)	$d_{IK,a}$	130,2	130,2	130,2
inner cryostat tube (ID)	$d_{IK,i}$	125,0	125,0	125,0
neutral conductor (OD)	$d_{NL,a}$	97,9	101,6	111,1
HTS screen (OD)	$d_{S,a}$	93,9	97,6	107,1
electrical isolation (OD)	$d_{EI,a}$	93,5	97,2	106,7
HTS conductor (OD)	$d_{L,a}$	41,8	48,9	58,2
hollow conductor (OD)	$d_{HL,a}$	41,0	48,1	57,4
former (OD)	$d_{T,a}$	36,0	44,0	54,0
former (ID)	$d_{T,i}$	32,0	40,0	50,0

Figure 3.1: Schematic representation in quarter cut and diameter of the examined cables with inner tube (hollow core conductor concept). (ID – Inner diameter, OD – Outer diameter).

Equation 3.1 also applies to the electrical insulation of the corrugated tube concept. The results of the design of the electrical insulation of the corrugated tube concept are shown in Table 3.3.

Table 3.3: Overview of the electric field strength as a function of the inner tube diameter (corrugated tube concept) and the electric insulation thickness

Nominal diameter of inner tube	DN 32	DN 40	DN 50
Thickness of the electrical isolation (Tyvek™)	28.25 mm	25.5 mm	24.0 mm
Electrical field strength-external conducting layer	4.8 kV/mm	5.7 kV/mm	6.5 kV/mm
Electrical field strength-internal conducting layer	12.5 kV/mm	12.5 kV/mm	12.3 kV/mm

The quarter cut and the diameters of a conceptual single-core cable with corrugated tube concept are shown in Figure 3.2.

Other functional layers, such as pads and semiconductor layers, are neglect due to their low layer thicknesses. The dimensions shown form the basis for the hydraulic and thermal calculations carried out in the following section.

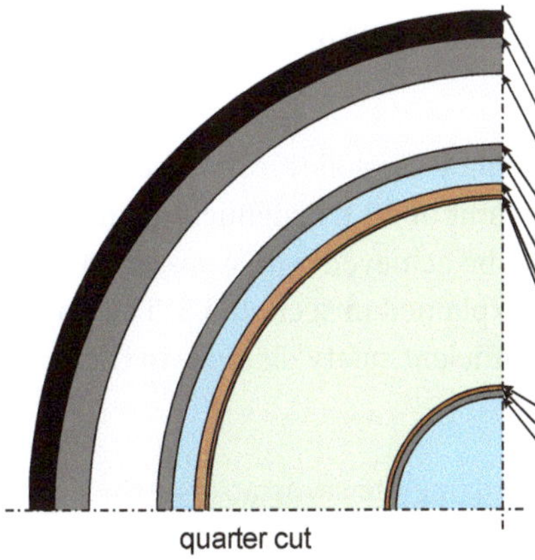

description	symbol	diameter in mm former nominal - DN 32	former nominal - DN 40	former nominal - DN 50
jacket (OD)	$d_{M,a}$	172,0	172,0	172,0
jacket (ID)	$d_{M,i}$	162,0	162,0	162,0
outer cryostat tube (ID)	$d_{AK,i}$	150,0	150,0	150,0
inner cryostat tube (OD)	$d_{IK,a}$	130,2	130,2	130,2
inner cryostat tube (ID)	$d_{IK,i}$	125,0	125,0	125,0
neutral conductor (OD)	$d_{NL,a}$	95,9	99,0	106,0
HTS screen (OD)	$d_{S,a}$	91,9	95,0	102,0
electrical isolation (OD)	$d_{EI,a}$	91,5	94,6	101,6
HTS conductor (OD)	$d_{L,a}$	35,0	43,6	53,6
former (OD)	$d_{IR,a}$	34,2	42,8	52,8
former (ID)	$d_{IR,i}$	32,0	40,0	50,0

Figure 3.2: Schematic representation in quarter cut and diameter of the examined cables with inner tube (corrugated tube concept). (ID – Inner diameter, OD – Outer diameter).

3.1.2 Temperature profile, mass flow and pressure drop

3.1.2.1 Operating range

The design of the superconducting cable includes hydraulic, electrical and thermal considerations. The operating range of the superconducting cable is decisive for the assessment of constructive influences on the three design criteria mentioned. The operating range must always be above the bubble-point curve of the liquid nitrogen so that no gas is formed. Gas formation must be avoided at all costs, otherwise electrical breakdown may occur. If the pressure falls below the bubble-point curve, the volume of nitrogen gas increases by a factor of 120 at a minimum operating pressure of 3 bar. When the temperature of the nitrogen gas reaches 300 K (ambient temperature, 3 bar), the volume expands by a factor of 800 [VDI13].

The critical point of nitrogen is -146.95°C (126.2 K) and 33.9 bar with a density of 0.314 g/cm^3. The area selected for this dimensioning remains way outside the super-critical area, which is set above these values and in which it is no longer possible to distinguish between liquid and gaseous nitrogen.

The maximum operating pressure depends on the permissible pressure in the former and in the inner cryostat tube. Typical values for this are 15-20 bar [CRY12]. If the permissible operating pressure is exceeded, a plastic longitudinal deformation of the corrugated tube occurs. The burst pressure must be according to standard "DIN EN 10380. Pipelines - Corrugated metal hoses and "metal hose assemblies" are four times higher than the maximum operating pressure [DIN80]. Further details on the corrugated tube follow in the section 3.1.2.3.

The natural lower limit for the temperature is given by the solidification point of nitrogen at 63.05 K. A practical lower temperature limit of 68 K is defined to ensure a sufficient distance from it. Such temperatures can be achieved with a so-called subcooler; its structure and functional principle are explained in section 3.5. The upper temperature limit is defined as 78 K to ensure a sufficient safety distance to the bubble-point curve.

The maximum operating pressure is thus achieved using the cryostats. For the operating range, however, a minimum operating pressure must also be defined. This results from the evaporation of the liquid nitrogen at too low pressure along the bubble-point curve. In order to ensure a safe distance from the evaporation at all times, the minimum operating pressure is set at 3 bar. Figure 3.3 shows the defined operating range and the bubble-point curve as a function of temperature and pressure.

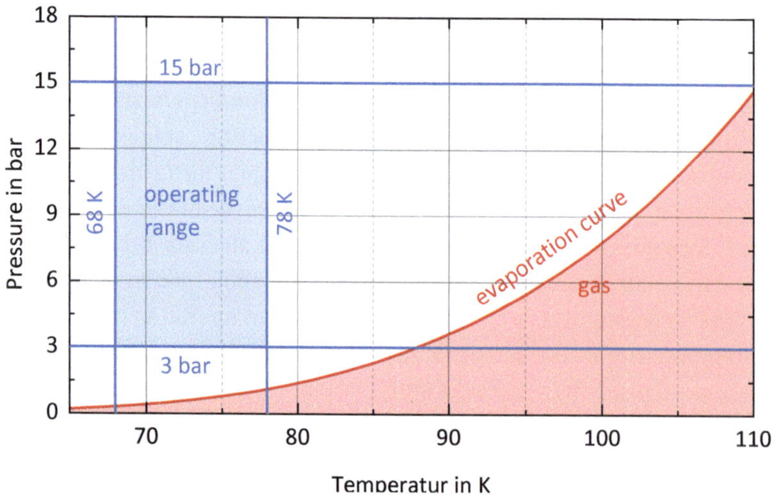

Figure 3.3: Representation of the defined operating range and the bubble-point curve of nitrogen as a function of temperature and pressure. The bubble-point curve separates the liquid (top) from the gaseous (bottom) state range of nitrogen.

A loss of pressure occurs over the length of a pipe or cryostats through which liquid flows. Due to the selected operating range for liquid nitrogen (maximum pressure 15 bar, minimum pressure 3 bar), this pressure drop must not exceed 12 bar over the entire length of the cable.

The mass flow rate is a fundamental parameter in the design of superconducting cables $\dot{m}$. This indicates how much mass nitrogen per time unit flows through the cable and can absorb thermal energy. Typical values for the mass flow $\dot{m}$ are 0.5 to 1.0 kg/s.

3.1.2.2 Cooling concepts

There are two different cooling concepts for single-core cables for three-phase applications: one-sided cooling and two-sided cooling. With one-sided cooling, the nitrogen is conducted in two cables. The return line is through a third cable, however with double mass flow. The result of this investigation is a maximum mass flow of $2\dot{m} = 2{,}0\ \mathrm{kg/s}$ for the return line in a single cable. Figure 3.4 represents the one-sided cooling schematically.

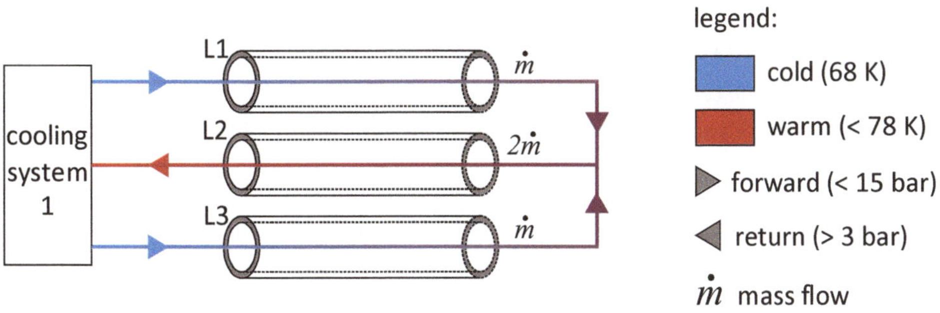

Figure 3.4: Schematic diagram of the one-sided cooling for a cable system with single-conductor cable.

For two-sided cooling, two chillers are used, which are placed at the two cable ends. The nitrogen is fed in at both ends of the cable at a temperature of 68 K and an operating pressure of 15 bar. This allows the entire operating range to be used for supply and return lines and allows greater route lengths than the one-sided coolant supply. The division of the mass flow into supply and return lines remains unchanged. The maximum hydraulic route length is thus determined by the return line with $2\dot{m}$, while the maximum thermal route length is limited by the supply line with $\dot{m}$. Figure 3.5 shows the two-sided cooling schematically.

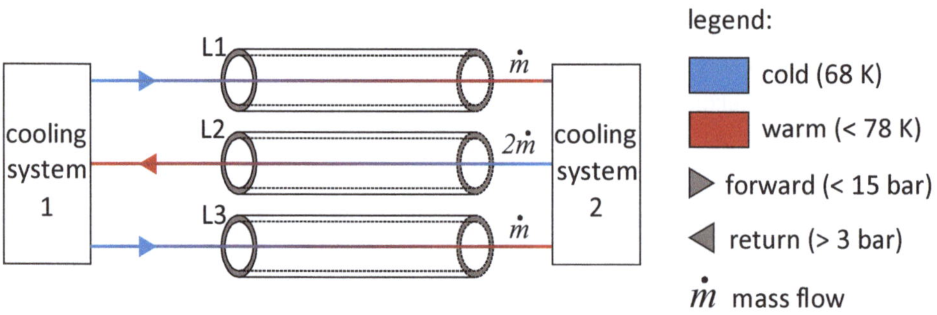

Figure 3.5: Schematic diagram of the two-sided cooling for a cable system with single-conductor cable.

3.1.2.3 Hydraulic consideration

The hydraulic consideration of the superconducting cable design investigates the pressure drop along the cable or former. The former fulfils two functions: First, it is used to feed in nitrogen. Second, it is used to transport part of the short-circuit current in the event of a fault.

This study examines two former concepts: the hollow core conductor concept and the corrugated tube concept. Previously completed projects with superconducting cable systems used corrugated stainless-steel tubes for the former construction. The advantage of a corrugated tube is its greater flexibility due to the small bending radius. In contrast to a simple corrugated tube, the spiral structure causes a lower pressure loss and a higher heat transfer coefficient [Kau71]. To produce the corrugated tube, a narrow stainless-steel sheet is bent into a tube, then longitudinally welded and rolled into the desired spiral-shaped corrugation. This manufacturing process is relatively simple and therefore inexpensive. This former concept is not suitable for compensating high short-circuit currents in the event of a fault.

In this study an additional copper layer is constructed around a carrier in order to increase the short-circuit current resistance. The spiral carrier has a thickness of 2 mm and is flexible so that it allows the cable to bend. The additional copper layer consists of copper wires, which together form a cross-sectional area of 300 mm^2. The thickness of the hollow core conductor, i.e. the thickness of the additional copper layer, varies depending on the carrier diameter. This design of the hollow core conductor concept is based on a conventional oil pressure cable [ESE75]. Further details on the short-circuit calculation and the influence of copper stabilisation are given in the section 3.3.

The standard "DIN EN 10380. Pipelines - Corrugated metal hoses and hose assemblies" defines the diameters of corrugated tubes with a nominal diameter of DN 20 to DN 300 [DIN80]. In this study, the carrier diameters of the hollow core conductor concept are related to these nominal diameters in addition to the spiral corrugated pipes.

Table 3.4 shows the dimensions of a hollow core conductor additionally stabilized with copper.

Table 3.4: Dimensions of the hollow core conductor with carrier for standardized nominal sizes

Nominal diameter	Inner diameter	Outside diameter	Hollow conductor strength	Carrier strength	Surface
	d_i	d_a	t	s	A_W
	mm	mm	mm	mm	m^2/m
32	32.0	41.0	2.5	2	0.1125
40	40.0	48.1	2.05	2	0.1336
50	50.0	57.4	1.7	2	0.1673

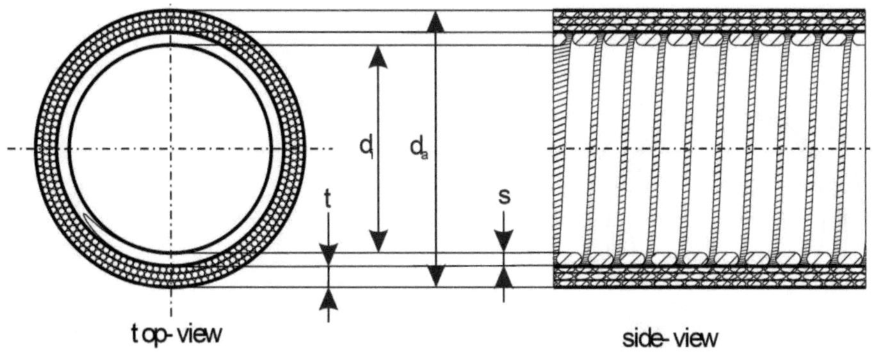

Figure 3.6: Schematic representation of an inner tube (hollow core conductor concept).

A schematic representation of the hollow core conductor concept is provided by Figure 3.6.

In the following section, the maximum nominal diameter for the corrugated tubes is limited to DN 150. More detailed information on materials suitable for low temperatures can be found in "DIN EN 13480-2 - Metallic industrial pipelines" [DIN48].

The dimensions of the spiral corrugated tubes for standardized nominal diameters are listed in Table 3.5 .

Table 3.5: The dimensions of the spiral corrugated tubes for standardized nominal diameters and the inner cryostat wall [DIN80].

Nominal diameter	Inner diameter	Outside diameter	Shaft distance	Corrugation depth	Wall thickness	Surface
	d_i	d_a	T	t	s	A_W
	mm	mm	mm	mm	mm	m^2/m
20	20.0	22.0	4.8	1.0	0.3	0.0686
25	25.0	27.2	5.3	1.1	0.3	0.0858
32	32.0	34.2	5.6	1.1	0.4	0.1125
40	40.0	42.8	6.8	1.4	0.5	0.1336
50	50.0	52.8	7.2	1.4	0.5	0.1673
65	65.0	68.4	8.6	1.7	0.6	0.2165
80	80.0	84.0	9.8	2.0	0.8	0.2635
100	100.0	104.6	11.4	2.3	0.8	0.3326
125	125.0	130.2	12.8	2.6	1.0	0.4148
150	150.0	156.0	15.0	3.0	1.0	0.5060

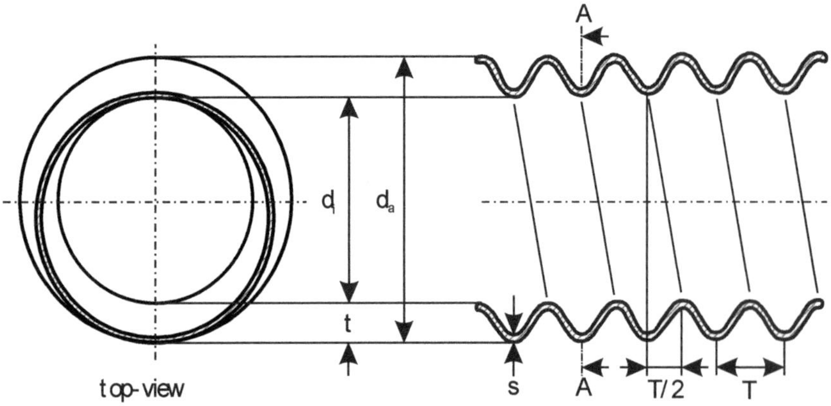

Figure 3.7: Schematic representation of a spiral corrugated pipe.

Figure 3.7 provides a schematic representation of a spiral-shaped corrugated tube.

In the literature [JHH15], there is no generally valid approach for the calculation of the pressure loss within a spiral corrugated pipe. In this study the pressure loss is calculated both for the hollow conductor concept and for the corrugated tube concept using the Darcy-Weisbach equation, as revised by Bernoulli [Kau70], for a plane smooth pipe according to equation 3.2 [Sha14]. The pipe friction coefficient λ_{IR} is determined according to Karman-Nikuradse in equation 3.3 [Det04].

$$\Delta p_{IR} = \frac{\lambda_{IR} \cdot l_{IR}}{d_{IR,i}} \cdot \frac{\rho_{LN}}{2} \cdot v_{IR}^2 \tag{3.2}$$

Required key indicators and their parameters:

$$Re_{IR} = \frac{v_{IR} \cdot d_{IR,i} \cdot \rho_{LN}}{\eta_{LN}} \tag{3.3}$$

$$\frac{1}{\lambda_R} = 2 \cdot log\left(\frac{Re_{IR} \cdot \sqrt{\lambda_R}}{2{,}51}\right) \tag{3.4}$$

$$v_{IR} = \frac{\dot{m}}{\rho_{LN} \cdot A_{IR}} = \frac{\dot{m}}{\rho_{LN} \cdot \left(\frac{\pi \cdot d_{IR,i}^2}{4}\right)} \tag{3.5}$$

Δp_{IR}	Pressure loss in the inner tube	(Pa)
Re_{IR}	Reynolds number in the inner tube	
λ_{IR}	Pipe friction coefficient	
l_{IR}	Length of the pipe	(m)
$d_{IR,i}$	Inner diameter of the inner tube	(m)
ρ_{LN}	Density of the nitrogen (VDI Heat Atlas)	(kg/m^3)
η_{LN}	Dynamic viscosity (VDI Heat Atlas)	($kg/m \cdot s$)
v_{IR}	Flow velocity of the liquid in the inner tube	(m/s)
$\dot{m}$	Mass flow	(kg/s)

The decisive variable for the pressure loss in a pipe is its diameter: The pressure loss falls with increasing diameter with the fifth power. In contrast, the pipe friction coefficient increases with increasing diameter, but this effect is comparatively small. With a quadratic dependence, the strength of the mass flow affects the pressure loss.

The maximum route length is calculated on the basis of the maximum permissible pressure drop of 12 bar. Table 3.6 indicates the maximum route lengths in dependency of the mass flow $\dot{m}$ for nominal value of the inner pipe from DN 32 to DN 50.

Table 3.6: Maximum hydraulic line length as a function of mass flow

Nominal diameter of inner pipe	Mass flow	Maximum route length in *m*	
		One-sided cooling	Double-sided cooling
DN 32	$\dot{m} = 0{,}5\,\mathrm{kg/s}$	2070	2666
	$\dot{m} = 1{,}0\,\mathrm{kg/s}$	590	758
DN 40	$\dot{m} = 0{,}5\,\mathrm{kg/s}$	6052	7793
	$\dot{m} = 1{,}0\,\mathrm{kg/s}$	1729	2223
DN 50	$\dot{m} = 0{,}5\,\mathrm{kg/s}$	>10000	>10000
	$\dot{m} = 1{,}0\,\mathrm{kg/s}$	5060	6508

It should be noted that the calculations are based on a horizontal route without height difference. A height profile can strongly influence the pressure curve over a route length. A height difference of approx. 150 m already causes a hydrostatic pressure of 12 bar and thus exceeds the permissible operating range. The pressure curve over a route length according to equation 3.6 is composed of the dynamic pressure and the hydrostatic pressure.

$$\begin{aligned} \Delta p_{\text{ges}}(\Delta z, \Delta h) &= \underbrace{\Delta p_{\text{IR}}}_{\text{Staudruck}} \pm \underbrace{\Delta p_{\text{HS}}}_{\text{hydrostatischer Druck}} \\ &= \frac{\lambda_{\text{IR}} \cdot v_{\text{IR}}^2}{d_{\text{IR,i}}} \cdot \frac{\rho_{\text{LN}}}{2} \cdot \Delta z \pm g \cdot \rho_{\text{LN}} \cdot \Delta h \end{aligned} \tag{3.6}$$

Δp_{ges}	Pressure loss (total)	(Pa)
Δp_{IR}	Pressure loss (inner tube - dynamic pressure)	(Pa)
Δp_{HS}	Pressure loss (inner tube - hydrostatic pressure)	(Pa)
g	Free-fall acceleration – 9.81 m/s²	(m/s²)
Δh	Differences in heigh	(m)
Δz	Differences in length	(m)

Equation 3.6 shows a linear relation between hydrostatic pressure and height difference. Due to the high sensitivity of the pressure curve to a height difference, the pressure curve above the given height profile must always be investigated in a project-related design.

3.1.2.4 Pump capacity

A pump is required to generate the desired mass flow $\dot{m}$. The pump capacity can be calculated with equation 3.7.

$$P_{\mathrm{PU}} = \frac{\dot{V} \cdot \Delta p_{\mathrm{ges}}}{\eta_{\mathrm{ges}}} = \frac{\dot{m} \cdot \Delta p_{\mathrm{ges}}}{\rho_{\mathrm{LN}} \cdot \eta_{\mathrm{ges}}} \tag{3.7}$$

P_{PU}	Pump capacity	(W)
η_{ges}	Efficiency (pump with pump motor), assumed at 0.7	

3.1.2.5 Electrical consideration

In addition to checking the design field strengths with equation 3.1, the electrical consideration also includes determining the current carrying capacity. An important factor in the electrical design is the number of HTS SC tapes running in parallel in the cable. A distinction is made between geometrically required SC tapes and electrically required SC tapes. The number of geometrically required SC tapes depends on the width of the HTS SC tapes and the dimensions of the cable core. For a given diameter or circumference, therefore, a certain number of SC tapes must be wound up. The number of electrically required SC tapes guarantees the current carrying capacity of the individual electrical phases. Accordingly, the geometrically required number of SC tapes must always be equal to or greater than the electrically required number of SC tapes:

$$N_{\mathrm{BL,geometrisch}} \geq N_{\mathrm{BL,elektrisch}}$$

If this condition is not met, the current carrying capacity is exceeded. The required number of SC tapes in each phase can be calculated according to equation 3.8 as a function of the critical current.

$$N_{\mathrm{BL,ges}} = \frac{\hat{\imath}_{\mathrm{r,max}}}{k_{\mathrm{SI}} \cdot I_{\mathrm{C}}} = \frac{I_{\mathrm{r,max}} \cdot \sqrt{2}}{k_{\mathrm{SI}} \cdot I_{\mathrm{C}}} \tag{3.8}$$

$N_{\mathrm{BL,ges}}$	Number of SC tapes per phase	
$\hat{\imath}_{\mathrm{r,max}}$	Peak value of the rated current	(A)
$I_{\mathrm{r,max}}$	RMS value of the rated current	(A)
I_{C}	Critical current of a SC tape	(A)
k_{SI}	Safety factor (maximum utilization of the SC tapes up to I_{C})	

In electrical power supply networks there may be brief current surges. A certain safety factor should be included in the layout of the cable to ensure that the SC tape does not lose its superconducting state when the current is increased, i.e. it does not go into the so-called quench. The safety factor k_{SI} is individually adapted to the respective cable system and, according to experience, is in the range of 0.5 to 0.8. This layout is independent of suitable protective measures, such as a short-circuit tolerant cable design or an external current limiter, which must additionally be installed in case of a short-circuit current. For more information, see section 3.3.

The minimum number of SC tapes $N_{\mathrm{BL,elektrisch}}$, that are wound in the cable results from the rated current for the cable based on the current carrying capacity of the conductor and the relevant safety factor. For the current carrying capacity of the conductor, it should be noted that the conductor is influenced by the magnetic field in the cable and therefore the value valid for each conductor individually - without an external field - is not reached in the network. This study assumes a critical current I_C of 150 A, which already includes the influence of the magnetic field.

The second criterion for determining the minimum number is the complete geometric coverage of the individual phases by HTS SC tapes. If this coverage is not provided and there are larger gaps between the individual strips, the AC losses in the cable will increase sharply due to the magnetic fields. Therefore, with three-phase concentric design, this geometric factor can become the determining factor for the design.

Table 3.7: Overview of the geometric and electrical number of SC tapes (per phase) as a function of the nominal width of the inner tube (hollow core conductor concept)

Nominal diameter of inner tube	Geometric number of SC tapes Layer 1	Geometric number of SC tapes Layer 2	Geometric number of SC tapes Total	Electrical number of SC tapes (Minimum)
	$N_{\mathrm{BL,1,geometrisch}}$	$N_{\mathrm{BL,1,geometrisch}}$	$N_{\mathrm{BL,geometrisch}}$	$N_{\mathrm{BL,elektrisch}}$
32	31	31	62	43
40	26	37	73	43
50	44	44	88	43

Note: Safety factor $k_{\mathrm{SI}} = 0{,}8$; SC tape width 4 mm; height 0.2 mm; critical current 150 A; rated current (effective) 3.6 kA

With an inner tube diameter of DN 50, a SC tape layer can be dispensed with, as the geometric number of SC tapes is greater than the electrical number of SC tapes with just one layer.

Table 3.8: Overview of the geometric and electrical number of SC tapes (per phase) as a function of the nominal diameter of the inner tube (corrugated tube concept)

Nominal diameter of inner tube	Geometric number of SC tapes Layer 1	Geometric number of SC tapes Layer 2	Geometric number of SC tapes Total	Electrical number of SC tapes (Minimum)
	$N_{BL,1,geometrisch}$	$N_{BL,1,geometrisch}$	$N_{BL,geometrisch}$	$N_{BL,elektrisch}$
32	26	26	52	43
40	32	33	65	43
50	40	40	80	43

Note: Safety factor $k_{SI} = 0{,}8$; SC tape width 4 mm; height 0.2 mm; critical current 150 A; rated current (effective) 3.6 kA

3.1.2.6 Thermal consideration

When designing the cable, the investigation of the heating of the cooling medium over the entire length of the cable route plays an important role. The heating of the liquid nitrogen is due to the electrical losses - dielectric losses and alternating current loss - of the cable inside the line cryostat and to a heat input from outside. For safe operation within the operating range shown in Figure 3.3, the maximum temperature of 78 K must not be exceeded at any point on the cable. For this it is necessary to calculate the temperature profile in radial and axial cable direction. The radial temperature curve can be calculated in layers by a group of Laplace differential equations. Equation 3.9 applies to solid layers with heat conduction, such as the inner tube and the neutral conductor.

$$\frac{1}{r}\frac{d}{dr}\left(r\frac{dT}{dr}\right) = 0 \tag{3.9}$$

If the solid layer also has a heat source $\dot{\omega}$ with thermal conductivity λ, such as the superconductor layer and the electrical insulation, the Poisson equation 3.10 is used.

$$\frac{d^2T}{dr^2} + \frac{1}{r}\cdot\frac{dT}{dr} + \frac{\dot{\omega}}{\lambda} = 0 \tag{3.10}$$

A group of differential equations which can be solved numerically and analytically are available for the thermal calculation of the hollow core conductor.

Layer identification	Differential equation	
Carrier	$\frac{1}{r}\frac{d}{dr}\left(r\frac{dT_{\mathrm{T}}}{dr}\right) = 0$	(3.16)
Hollow core conductor	$\frac{1}{r}\frac{d}{dr}\left(r\frac{dT_{\mathrm{HL}}}{dr}\right) = 0$	(3.17)
HTS conductor	$\frac{d^2T_{\mathrm{L}}}{dr^2} + \frac{1}{r}\cdot\frac{dT_{\mathrm{L}}}{dr} + \frac{\dot{\omega}_{\mathrm{L}}}{\lambda_{\mathrm{HTSL}}} = 0$	(3.18)
Electrical isolation	$\frac{d^2T_{\mathrm{EI}}}{dr^2} + \frac{1}{r}\cdot\frac{dT_{\mathrm{EI}}}{dr} + \frac{\dot{\omega}_{\mathrm{EI}}}{\lambda_{\mathrm{TY}}} = 0$	(3.19)
HTS shield	$\frac{d^2T_{\mathrm{S}}}{dr^2} + \frac{1}{r}\cdot\frac{dT_{\mathrm{S}}}{dr} + \frac{\dot{\omega}_{\mathrm{S}}}{\lambda_{\mathrm{HTSL}}} = 0$	(3.20)
Neutral conductor	$\frac{1}{r}\frac{d}{dr}\left(r\frac{dT_{\mathrm{NL}}}{dr}\right) = 0$	(3.21)

T_{T}	Temperature curve in the carrier	(K)
T_{HL}	Temperature curve in the hollow core conductor	(K)
T_{EI}	Temperature profile in the electrical insulation	(K)
T_{S}	Temperature curve in the shield	(K)
T_{NL}	Temperature curve in the neutral conductor	(K)
λ_{HTSL}	Thermal conductivity of the HTS SC tapes (YBCO)	(W/m · K)
λ_{TY}	Thermal conductivity of electrical insulation (Tyvek™)	(W/m · K)

In this case, twelve boundary conditions are required to determine the special solution.

RB 1	$\lambda_{\mathrm{FE}} \cdot \frac{d}{dr}\left(T_{\mathrm{T}}\left(r_{\mathrm{T,in}}\right)\right) = \alpha_{\mathrm{T}}[T_{\mathrm{T}}(r_{\mathrm{T}}) - T_{\mathrm{LN}}(z)]$
RB 2	$\lambda_{\mathrm{FE}} \cdot \frac{d}{dr}\left(T_{\mathrm{T}}\left(r_{\mathrm{IR,au}}\right)\right) = \lambda_{\mathrm{CU}} \cdot \frac{d}{dr}\left(T_{\mathrm{HL}}\left(r_{\mathrm{HL,in}}\right)\right)$
RB 3	$T_{\mathrm{T}}\left(r_{\mathrm{T,au}}\right) = T_{\mathrm{HL}}\left(r_{\mathrm{HL,in}}\right)$
RB 4	$\lambda_{\mathrm{CU}} \cdot \frac{d}{dr}\left(T_{\mathrm{HL}}\left(r_{\mathrm{HL,au}}\right)\right) = \lambda_{\mathrm{HTSL}} \cdot \frac{d}{dr}\left(T_{\mathrm{L}}\left(r_{\mathrm{L,in}}\right)\right)$
RB 5	$T_{\mathrm{HL}}\left(r_{\mathrm{HL,au}}\right) = T_{\mathrm{L}}\left(r_{\mathrm{L,in}}\right)$
RB 6	$\lambda_{\mathrm{HTSL}} \cdot \frac{d}{dr}\left(T_{\mathrm{L}}\left(r_{\mathrm{L,au}}\right)\right) = \lambda_{\mathrm{TY}} \cdot \frac{d}{dr}\left(T_{\mathrm{EI}}\left(r_{\mathrm{EI,in}}\right)\right)$
RB 7	$T_{\mathrm{L}}\left(r_{\mathrm{L,au}}\right) = T_{\mathrm{EI}}\left(r_{\mathrm{EI,in}}\right)$

RB 8	$\lambda_{TY} \cdot \frac{d}{dr}\left(T_{EI}\left(r_{EI,au}\right)\right) = \lambda_{HTSL} \cdot \frac{d}{dr}\left(T_S\left(r_{S,in}\right)\right)$
RB 9	$T_{EI}\left(r_{EI,au}\right) = T_S\left(r_{S,in}\right)$
RB 10	$\lambda_{HTSL} \cdot \frac{d}{dr}\left(T_S\left(r_{S,au}\right)\right) = \lambda_{CU} \cdot \frac{d}{dr}\left(T_{NL}\left(r_{NL,in}\right)\right)$
RB 11	$T_S\left(r_{S,au}\right) = T_{NL}\left(r_{NL,in}\right)$
RB 12	$\lambda_{CU} \cdot \frac{d}{dr}\left(T_{NL}\left(r_{NL,au}\right)\right) = \dot{q}_{rad}$

Required key indicators and their parameters:

λ_{FE}	Thermal conductivity of the inner tube wall (stainless steel) 1.5662)	$(W/m \cdot K)$
λ_{CU}	Thermal conductivity of the neutral conductor (copper)	$(W/m \cdot K)$
α_{IR}	Heat transfer coefficient	(W/m^2K)
$\dot{q}_{rad}$	Heat input through cable cryostats	(W/m^2)

The temperature curves for the corrugated pipe concept are calculated taking into account the changed layer structure.

Layer identification	Differential equation	
Inner tube	$\frac{1}{r}\frac{d}{dr}\left(r\frac{dT_{IR}}{dr}\right) = 0$	(3.11)
HTS conductor	$\frac{d^2T_L}{dr^2} + \frac{1}{r} \cdot \frac{dT_L}{dr} + \frac{\dot{\omega}_L}{\lambda_{HTSL}} = 0$	(3.12)
Electrical isolation	$\frac{d^2T_{EI}}{dr^2} + \frac{1}{r} \cdot \frac{dT_{EI}}{dr} + \frac{\dot{\omega}_{EI}}{\lambda_{TY}} = 0$	(3.13)
HTS shield	$\frac{d^2T_S}{dr^2} + \frac{1}{r} \cdot \frac{dT_S}{dr} + \frac{\dot{\omega}_S}{\lambda_{HTSL}} = 0$	(3.14)
Neutral conductor	$\frac{1}{r}\frac{d}{dr}\left(r\frac{dT_{NL}}{dr}\right) = 0$	(3.15)

T_{IR}	Temperature curve in the inner tube	(K)
T_L	Temperature curve in the HTS conducto	(K)
T_{EI}	Temperature profile in the electrical insulation	(K)
T_S	Temperature curve in the shield	(K)
T_{NL}	Temperature curve in the neutral conductor	(K)
λ_{HTSL}	Thermal conductivity of the HTS SC tapes (YBCO)	$(W/m \cdot K)$

λ_{TY}	Thermal conductivity of electrical insulation (Tyvek™)	$(W/m \cdot K)$
$\dot{\omega}_L$	Heat source - AC losses of the HTS conductors	(W/m^3)
$\dot{\omega}_{EI}$	Heat source - dielectric losses of electrical insulation	(W/m^3)
$\dot{\omega}_S$	Heat source - AC losses of the HTS shield	(W/m^3)

The general solution to the differential equations is given in the Annex. In this case, ten boundary conditions are required to determine the special solution.

RB 1	$\lambda_{FE} \cdot \frac{d}{dr}\left(T_{IR}(r_{IR,in})\right) = \alpha_{IR}[T_{IR}(r_{IR}) - T_{LN}(z)]$
RB 2	$\lambda_{FE} \cdot \frac{d}{dr}\left(T_{IR}(r_{IR,au})\right) = \lambda_{HTSL} \cdot \frac{d}{dr}\left(T_L(r_{L,in})\right)$
RB 3	$T_{IR}(r_{IR,au}) = T_L(r_{L,in})$
RB 4	$\lambda_{HTSL} \cdot \frac{d}{dr}\left(T_L(r_{L,au})\right) = \lambda_{TY} \cdot \frac{d}{dr}\left(T_{EI}(r_{EI,in})\right)$
RB 5	$T_L(r_{S,au}) = T_{EI}(r_{EI,in})$
RB 6	$\lambda_{TY} \cdot \frac{d}{dr}\left(T_{EI}(r_{EI,au})\right) = \lambda_{HTSL} \cdot \frac{d}{dr}\left(T_S(r_{S,in})\right)$
RB 7	$T_{EI}(r_{EI,au}) = T_S(r_{S,in})$
RB 8	$\lambda_{HTSL} \cdot \frac{d}{dr}\left(T_S(r_{S,au})\right) = \lambda_{CU} \cdot \frac{d}{dr}\left(T_{NL}(r_{NL,in})\right)$
RB 9	$T_S(r_{S,au}) = T_{NL}(r_{NL,in})$
RB 10	$\lambda_{NL} \cdot \frac{d}{dr}\left(T_{NL}(r_{NL,au})\right) = \dot{q}_{rad}$

Required key indicators and their parameters:

λ_{FE}	Thermal conductivity of the inner tube wall (stainless steel) 1.5662)	$(W/m \cdot K)$
λ_{CU}	Thermal conductivity of the neutral conductor (copper)	$(W/m \cdot K)$
α_{IR}	Heat transfer coefficient	(W/m^2K)
$\dot{q}_{rad}$	Heat input through cable cryostats	(W/m^2)

Table 3.9 indicates the thermal conductivity of the cable materials in the average temperature range from 68 K to 78 K.

Table 3.9: Thermal conductivity of the various cable materials as an average value between 68 K and 78 K

Material	Thermal conductivity Symbol	Value (W/m · K)
Copper	λ_{CU}	576
Stainless steel	λ_{FE}	7.8
YBCO SC tape	λ_{HTSL}	150
PPLP	λ_{PP}	0.05
Tyvek™ (Polyethylene)	λ_{TY}	0.3
Nitrogen (liquid)	λ_{LN}	0.15

Equations 3.16 and 3.17 are used to calculate the axial temperature increase [Sha17[1]].

$$\Delta T = \frac{\alpha_{\mathrm{IR}} \cdot A_{\mathrm{W}} \cdot \left(T(r_{\mathrm{IR,in}}) - T_{\mathrm{LN}}(z)\right)}{\dot{m} \cdot c_{\mathrm{p,LN}}} \cdot \Delta z \tag{3.16}$$

$$T_{\mathrm{LN}}(z + \Delta z) = \Delta T + T_{LN}(z) \tag{3.17}$$

ΔT	Axial temperature increase per step size Δz	(K)
Δz	Axial increment	(m)
$T_{\mathrm{LN}}(z + \Delta z)$	Axial nitrogen temperature	(K)
α_{IR}	Heat transfer coefficient between nitrogen and inner tube	(W/m^{2}K)
A_{W}	Surface of the inner tube (Table 3.5)	(m^2/m)
$T(r_{\mathrm{IR,in}})$	Temperature on the inside of the inner tube	(K)
$T_{\mathrm{LN}}(z)$	Temperature in nitrogen at the inner tube	(K)
$\dot{m}$	Mass flow of nitrogen	(kg/s)
$c_{\mathrm{p,LN}}$	Specific heat capacity (VDI Heat Atlas)	(J/kg · K)

The heat transfer coefficient α_{IR} and the heat input $\dot{q}_{\mathrm{rad}}$ are calculated in the following sections.

Convective heat transport in the inner tube

Convective heat transfer can be described with the heat transfer coefficient α_{IR}. This determines the intensity of the heat transfer at an interface, in this case at the inter-

[1] http://dx.doi.org/10.1016/j.cryogenics.2016.11.004

face between the inner tube and the liquid nitrogen. The calculation of the heat transfer coefficient for the inner pipe according to equation 3.18 is carried out using several dimensionless parameters.

$$\alpha_{\mathrm{IR}} = \frac{Nu_{\mathrm{IR}} \cdot \lambda_{\mathrm{LN}}}{d_{\mathrm{IR,in}}} \tag{3.18}$$

Dimensionless key indicators:

$$Nu_{\mathrm{IR}} = \frac{\left(\frac{\lambda_R}{8}\right) \cdot Re_{\mathrm{IR}} \cdot Pr}{1 + 12{,}7 \cdot \sqrt{\frac{\lambda_R}{8}} \cdot \left(Pr^{\frac{2}{3}} - 1\right)} \cdot K_{\mathrm{L}} \tag{3.19}$$

$$Pr = \frac{\eta_{\mathrm{LN}} \cdot c_{\mathrm{p,LN}}}{\lambda_{\mathrm{LN}}} \tag{3.20}$$

$$K_{\mathrm{L}} = \left(1 + \left(\frac{d_{\mathrm{IR,in}}}{l_{\mathrm{IR}}}\right)^{\frac{2}{3}}\right) \tag{3.21}$$

$$\lambda_{\mathrm{R}} = (1{,}8 \cdot log_{10}(Re) - 1{,}5)^{-2} \tag{3.22}$$

α_{IR}	Heat transfer coefficient between nitrogen and inner tube	$(\mathrm{W/m^2K})$
Nu_{IR}	Nusselt number	
Re_{IR}	Reynolds number (equation 3.2)	
Pr	Prandtl number	
K_L	Length correction factor	
λ_R	Pipe friction coefficient	
λ_{LN}	Thermal conductivity (VDI Heat Atlas)	$(\mathrm{W/m \cdot K})$
η_{LN}	Dynamic viscosity (VDI Heat Atlas)	$(\mathrm{kg/m \cdot s})$
$c_{\mathrm{p,LN}}$	Specific heat capacity (VDI Heat Atlas)	$(\mathrm{J/kg \cdot K})$
$d_{\mathrm{IR,in}}$	Diameter of the inner tube	(m)
l_{IR}	Length of the inner tube	(m)

Heat source - heat input through the line cryostat

The thermal conductivity of the gas, the contact point heat conduction and the heat transmission by radiation must be taken into account when calculating the heat input from the outside. The specific heat input is given in the manufacturer's data [CRY12]. The values taken from this for the heat input $P'_{\mathrm{V,0,LK}}$ are listed in Table 3.10. The values for nominal diameter greater than DN 80 were extrapolated.

Table 3.10: Specific heat input from CRYOFLEX transfer pipes [CRY12]

Nominal diameter inner cryostat tube	Heat radiation	
	W/m	W/m²
DN 20	0.8	11.7
DN 25	1.2	13.9
DN 32	1.5	13.3
DN 40	1.6	11.9
DN 50	1.7	10.2
DN 65	1.9	8.8
DN 80	2.0	7.6
DN 100	2.2	6.6
DN 125	2.3	5.6
DN 150	2.5	4.9

Heat source - AC losses in the conductor and shielding layers

The temperature-dependent alternating current losses in SC tape of the superconducting cable are required to calculate the internal heat source $\dot{\omega}$. The detailed examination of alternating current losses shows that three types of loss can be distinguished: Coupling losses, eddy current losses and hysteresis losses.

The coupling losses are suppressed by the twisting of the conductors; they are therefore neglected in the following. The eddy current losses are not considered in this study, because they are only significant from a frequency of 1 kHz. Therefore, only the hysteresis losses have to be considered. These are caused by the fact that the superconductors inside are partially traversed by magnetic fluxes. These can be self-fields or external fields. This only happens at certain points in the material, at the so-called flux lines or vortices. If an electric current flows through the superconductor, the Lorentz force causes a displacement of the flux lines. This movement generates thermal energy in the material, which represents a loss and must be negated by the cooling.

A simple model for estimating hysteresis losses of circularly arranged SC tapes in superconducting cables, uses the approximation equation 3.23 [Nor70[2]].

$$Q'_{\mathrm{V,L}}(T) = \frac{\mu_0 I_{\mathrm{C}}^2(T)}{\pi}\left[\left(1 - F_{\mathrm{i}}(T)\right)ln\left(1 - F_{\mathrm{i}}(T)\right) + \left(1 + F_{\mathrm{i}}(T)\right)ln\left(1 + F_{\mathrm{i}}(T)\right) - F_{\mathrm{i}}^2(T)\right] \quad (3.23)$$

[2] https://doi.org/10.1088/0022-3727/3/4/308

$$F_{\mathrm{i}}(T) = \frac{\hat{I}_{\mathrm{i}}}{I_{\mathrm{C}}(T)} = \frac{I_{\mathrm{N}} \cdot \sqrt{2}}{I_{\mathrm{C}}(T) \cdot N_{\mathrm{LS}}} \tag{3.24}$$

$Q'_{\mathrm{V,L}}$	Energy loss per cycle	(J/m)
μ_0	Magnetic field constants	($\mathrm{kgm/s^2A^2}$)
I_{C}	Critical current of SC tapes	(A)
F_{i}	Load factor of SC tapes	
$\hat{I}_{\mathrm{i}}$	Peak value of the current	(A)
$I_{\mathrm{C}}(T)$	Critical current per SC tape	(A)
I_{N}	Nominal current per phase (3,600 A)	(A)
N_{LS}	Number of conductor layers	

Equation 3.25 can be used to calculate the critical amperage under consideration of the temperature [SZ65].

$$I_{\mathrm{C}}(T) = I_{\mathrm{C}} \cdot \left(\frac{T_{\mathrm{C}} - T}{T_{\mathrm{C}} - T_0}\right)^a \tag{3.25}$$

$I_{\mathrm{C}}(T)$	Critical current as a function of temperature	(A)
I_{C}	Critical current of a SC tape 150 A	(A)
T_{C}	Critical temperature $T_{\mathrm{C}} = 90\ K$	(K)
T	Rated temperature	(K)
T_0	Temperature at the operating point of the SC tape	$T_0 = 77\ K$(K)
a	Exponent of temperature dependence	$a = 1{,}5$

With the equation 3.23 the energy loss $Q_{\mathrm{V,SL}}$ in a flat SC tape is calculated. The resulting power loss $P_{\mathrm{V,SL}}$ is calculated by multiplying the frequency f_{NZ} and the number of SC tapes per layer $N_{\mathrm{BL,i}}$ in equation 3.26.

$$P'_{\mathrm{V,L}}(T) = \sum_{i=1}^{N_{\mathrm{LS}}} Q_{\mathrm{V,L}}(T) \cdot N_{\mathrm{L,i}} \cdot f_{\mathrm{NZ}} \tag{3.26}$$

$P'_{\mathrm{V,L}}$	Total power loss	(W/m)
$N_{\mathrm{L,i}}$	Number of SC tapes per conductor layer	
f_{NZ}	Line frequency	(1/s)

The relation between internal heat source $\dot{\omega}$ and power loss is shown in equation 3.27.

$$\dot{\omega}_{\mathrm{L}}(T) = \frac{P_{\mathrm{V,L}}(T)}{A_{\mathrm{L}}} \tag{3.27}$$

$\dot{\omega}_{\mathrm{L}}$	Heat source - AC losses of the HTS conductor layer	$(\mathrm{W/m^3})$
A_{L}	Cross-sectional area of the SC tape layer	$(\mathrm{m^2})$

The calculation of the shield losses is analogous to the loss calculation of the conductor.

Heat source - dielectric losses of electrical insulation

The dielectric losses $P_{\mathrm{V,0,D}}$ of a cable are calculated according to equation 3.26 from the line frequency *f*, the capacitance per unit length *C'*, the loss factor $tan\delta$ and the nominal voltage U_{N}. The equation 3.28 of the coaxial cable is selected for the calculation of the capacitance.

$$P'_{\mathrm{V,0,D}} = 2\pi \cdot f_{\mathrm{NZ}} \cdot C' \cdot tan\delta \cdot U_{\mathrm{LE}}^2 \tag{3.28}$$

$$C' = \frac{2\pi \cdot \varepsilon_0 \cdot \varepsilon_{\mathrm{TY}}}{ln\left(\frac{r_{\mathrm{EI,au}}}{r_{\mathrm{EI,in}}}\right)} \tag{3.29}$$

$P'_{\mathrm{V,0,D}}$	Dielectric losses of the insulation	$(\mathrm{W/m})$
C'	Capacitance per unit length	$(\mathrm{F/m})$
$tan\delta$	Loss factor of insulation (1.3e-4 [HNM14])	
U_{LE}	Conductor-earth voltage $U_{\mathrm{LE}} = U_{\mathrm{N}}/\sqrt{3}$	(V)
ε_0	Electrical field constant	$(\mathrm{As/Vm})$
$\varepsilon_{\mathrm{TH}}$	Relative permittivity of the insulation (Tyvek™) (1.73 [HNM14])	

The correlation between the heat source $\dot{\omega}$ and the dielectric power loss is shown in the equation 3.30.

$$\dot{\omega}_{\mathrm{EI}} = \frac{P_{\mathrm{V,0,D}}}{A_{\mathrm{EI}}} \tag{3.30}$$

$\dot{\omega}_{\mathrm{EI}}$	Heat source - dielectric losses	$(\mathrm{W/m^3})$
A_{EI}	Cross-sectional area of the electrical insulation	$(\mathrm{m^2})$

3.1.3 Temperature profiles, design limits and operating points

The thermal model of the cable described in the section 3.1.2.6 allows the calculation of the axial and radial temperature profiles. Figure 3.8 shows, as example, the axial nitrogen temperature variation of the one-sided cooling for a nominal diameter of the inner pipe DN 40 over the route length (supply and return line).

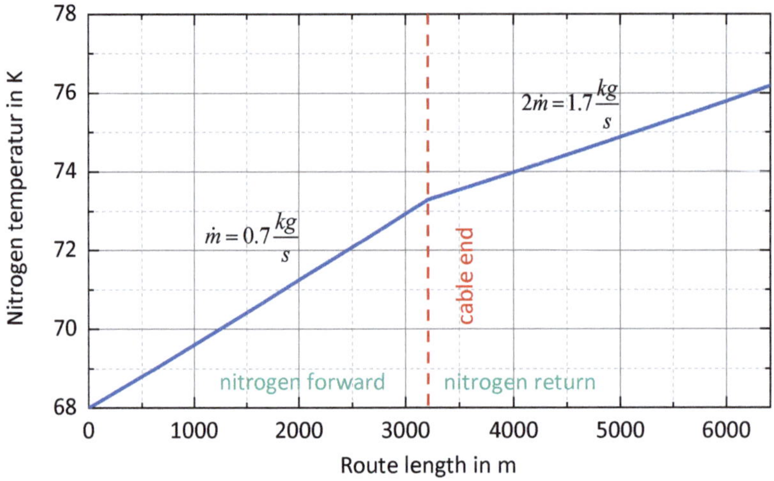

Figure 3.8: Axial nitrogen temperature curve of the one-sided cooling for a nominal diameter of the inner pipe DN 40 over the route length.

In Figure 3.8 two areas are visible, separated by the marking of the cable end. The temperature profile of the nitrogen introduction is in the left display range; the mass flow example is 0.7 kg/s. With one-sided cooling, the nitrogen from phase L1 and L3 is combined and returned to phase L2 with double mass flow 1.4 kg/s (see also Figure 3.4). This can be seen in the right display range. The nitrogen temperature at the coolant outlet is approx. 76 K.

Figure 3.9 shows the axial nitrogen temperature curve of the two-sided cooling for a nominal diameter of the inner tube DN 50 over the route length. The mass flow in phase L1 and L3 is, for example, 0.8 kg/s.

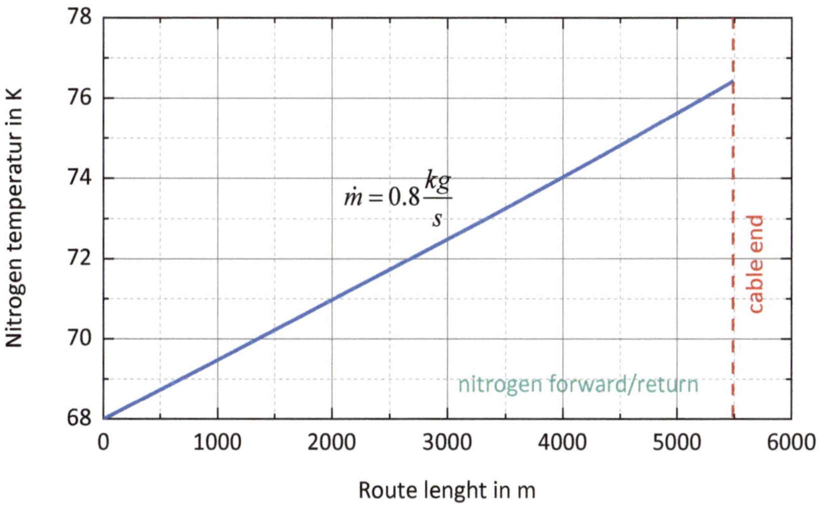

Figure 3.9: Axial nitrogen temperature curve of the two-sided cooling for a nominal diameter of the inner pipe DN 40 over the route length.

In the case of double-sided cooling the nitrogen in phase L1 and L3 is transported to the end of the cable and then conveyed back with double mass flow by phase L2. The nitrogen at the cable end is cooled down to 68 K by a second cooling system (see also Figure 3.5). Only the phase with the smallest mass flow (L1 or L3) is used to calculate the maximum thermal route length. This can be justified by the fact that with single mass flow in phase L1 or L3, the flow velocities and thus the convection coefficient are lower than in phase L2 with double mass flow. This results in poorer heat dissipation and higher cable temperatures. Therefore, this approach serves as a worst-case estimate.

The calculation of the axial nitrogen temperature is not sufficient to determine the maximum thermal route length. The thermal model also calculates the radial temperature distribution for this purpose. The radial temperature profile at the cable end for the corrugated pipe concept is shown in Figure 3.10.

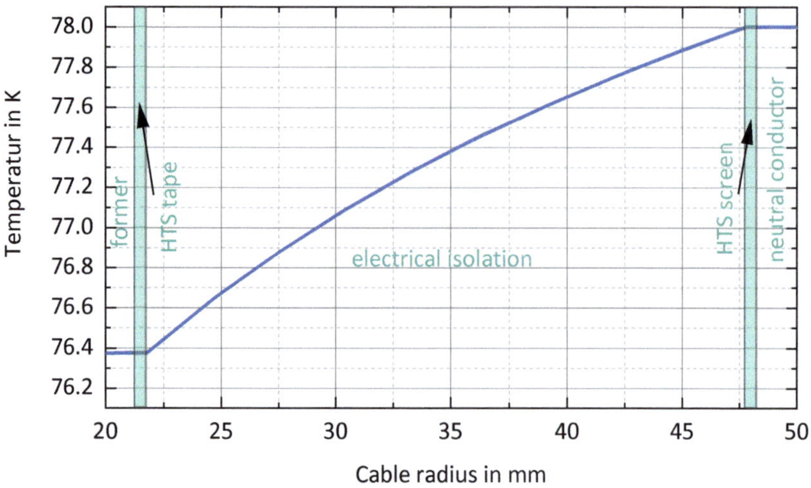

Figure 3.10: Radial temperature curve at cable end between inner tube and neutral conductor.

Figure 3.10 shows clearly the large impact of the electrical isolation, its thickness, its thermal conductivity and its dielectric losses have on the temperature calculation. At the fixed operating range of 10 K, the temperature difference across the insulation is already 1.5 K. In comparison to the temperature gradient in the electrical insulation, the temperature gradient in the inner pipe, in the HTS SC tape layers, the HTS shield layers and in the neutral conductor can become almost completely negligible.

Using axial and radial temperature calculations, it is possible to calculate the maximum thermal route length for a permissible temperature throughout the cable of < 78 K. Table 3.11 indicates the maximum route lengths in dependency of inside pipe diameter and mass flow for the hollow core conductor concept.

Table 3.11: Maximum thermal line lengths of the one- and two-sided cooling depending on the nominal width of the inner tube (hollow core conductor concept) and the mass flow

Nominal diameter of inner pipe	Mass flow	Maximum thermal route length in m	
		One-sided cooling	Double-sided cooling
DN 32	$\dot{m} = 0{,}5\,\mathrm{kg/s}$	2415	3630
	$\dot{m} = 1{,}0\,\mathrm{kg/s}$	4830	7250
DN 40	$\dot{m} = 0{,}5\,\mathrm{kg/s}$	2470	3710
	$\dot{m} = 1{,}0\,\mathrm{kg/s}$	4940	7410
DN 50	$\dot{m} = 0{,}5\,\mathrm{kg/s}$	2380	3570
	$\dot{m} = 1{,}0\,\mathrm{kg/s}$	4760	7140

A maximum of line lengths for the nominal diameter DN 40 of the inner tube for both cooling concepts results from Table 3.11. This maximum results from two influencing variables that work in opposite directions.

If the nominal diameters of the inner tube increase, the number of HTS SC tapes $N_{BL,1}/N_{BL,2}$ increases in the respective layer (1.2). This reduces AC losses $P_{V,SL}(l_{max})$. The cryostat transmission losses remain constant with the dimensions shown in Figure 3.1. The capacitance per unit length increases as the inner tube increases, the dielectric losses increase. Despite increasing dielectric losses, a lower total power loss is achieved by increasing the nominal diameter of the inner tube. The reduction of total losses leads to increasing maximum route lengths. The losses are specified in Table 3.12. The maximum power losses refer to the cable end.

Table 3.12: Overview of the different loss mechanisms as a function of the nominal diameter of the inner tube (hollow core conductor concept)

Label	Symbol	Unit	Nominal diameter of inner pipe		
			DN 32	DN 40	DN 50
AC conductor losses at cable end	$P'_{V,L}(l_{max})$	W/m	0.40	0.25	0.14
AC shielding losses at cable end	$P'_{V,S}(l_{max})$	W/m	0.26	0.24	0.18
Dielectric losses	$P'_{V,0,D}$	W/m	0.23	0.28	0.31
Cryostat transmission losses	$P'_{V,0,LK}$	W/m	2.30	2.30	2.30
Sum		W/m	3.19	3.07	2.93

The quantitative distribution of heat sources for the hollow core conductor concept is shown in the appendix. The cooling concept and the mass flow of the nitrogen have no influence on the AC losses at the end of the cable when considering the maximum route length.

In comparison to the reduction of the total losses, an enlargement of the inner tube leads to a reduced flow velocity v_{IR} of the nitrogen and thus to a decrease of the heat transfer coefficientα_{IR}. Reduced heat transfer reduces maximum route length. The discussed influencing variables are presented in detail in Table 3.13.

Table 3.13: Overview of the flow velocity of nitrogen and the heat transfer coefficient for the two cooling concepts as a function of the mass flow and of the nominal diameter of the inner tube.

	Symbol	Unit	One- and two-sided cooling Nominal diameter of inner pipe		
			DN 32	DN 40	DN 50
Mass flow	v_{IR}	m/s	1.54	0.98	0.63
$\dot{m} = 0{,}5\,kg/s$	α_{IR}	W/m^2K	3147	2086	1405
Mass flow	v_{IR}	m/s	3.07	1.97	1.26
$\dot{m} = 1{,}0\,kg/s$	α_{IR}	W/m^2K	5637	3731	2507

This means that any enlargement of the inner pipe at constant mass flow limits does not necessarily lead to an extension of the route length. A similar picture emerges for the corrugated tube concept, as Table 3.14 shows.

Table 3.14: Maximum thermal line lengths of the one- and two-sided cooling depending on the nominal width of the inner tube (hollow core conductor concept) and the mass flow

Nominal diameter of inner pipe	Mass flow	Maximum thermal route length in m	
		One-sided cooling	Double-sided cooling
DN 32	$\dot{m} = 0{,}5\,kg/s$	2015	3020
	$\dot{m} = 1{,}0\,kg/s$	4030	6050
DN 40	$\dot{m} = 0{,}5\,kg/s$	2290	3440
	$\dot{m} = 1{,}0\,kg/s$	4585	6880
DN 50	$\dot{m} = 0{,}5\,kg/s$	2285	3430
	$\dot{m} = 1{,}0\,kg/s$	4575	6860

In the following section, the maximum hydraulic route length (Table 3.6) and the maximum thermal route length (Table 3.11 or Table 3.14) are considered as a function of the mass flow. This provides an overview of the maximum route length depending on the mass flow, the nominal diameter of the inner tube and the cooling concept. In Figure 3.11 the maximum route length is shown as example based on the mass flow with one-sided cooling for the nominal diameter DN 40 of the inner pipe.

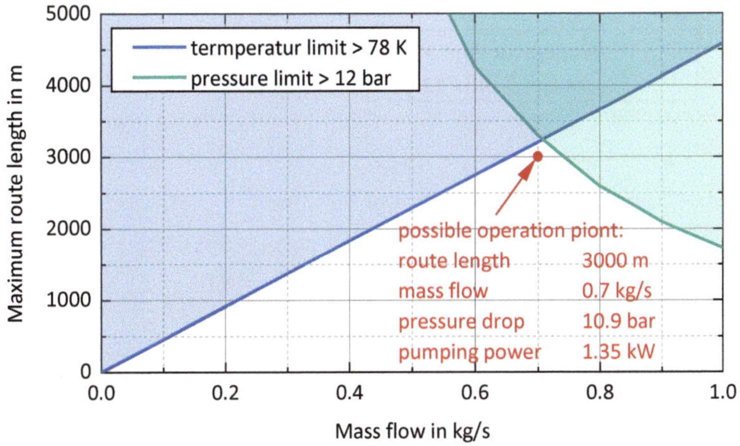

Figure 3.11: Representation of the maximum route length as a function of the mass flow with one-sided cooling for the nominal width DN 40 of the inner tube.

The area with the possible operating points lies below the thermal (blue) and hydraulic (green) boundaries. An operating point with a route length of 3000 m with mass flow, pressure loss and pump capacity is highlighted as an example. The minimum mass flow for a given route length can also be determined on the basis of Figure 3.11.

Figure 3.12 shows the correlation between maximum line length and mass flow with two-sided cooling with a nominal diameter DN 50 of the inner tube. It can be seen that the choice of the larger inside tube width shifts the hydraulic boundary in direction of higher route lengths. The use of the two-sided cooling concept also increases the maximum route length.

An almost linear dependency between maximum route length and mass flow of nitrogen can be determined for both cooling concepts.

For one-sided and two-sided cooling, the maximum line lengths are shown in the attachment A as a function of the mass flow and of the temperature and pressure limits for the nominal sizes DN 32 to DN 50 (see also section 1, Introduction). The maximum route lengths are on average 3-5 km; this enables typical partial cabling lengths in this range.

Reduction of the losses of line cryostats

One of the largest influencing factors on the maximum route length are the cryostat transmission losses according to Table 3.10. Theoretical calculations of these losses lead to much lower values than those specified by the manufacturer [Sha14].

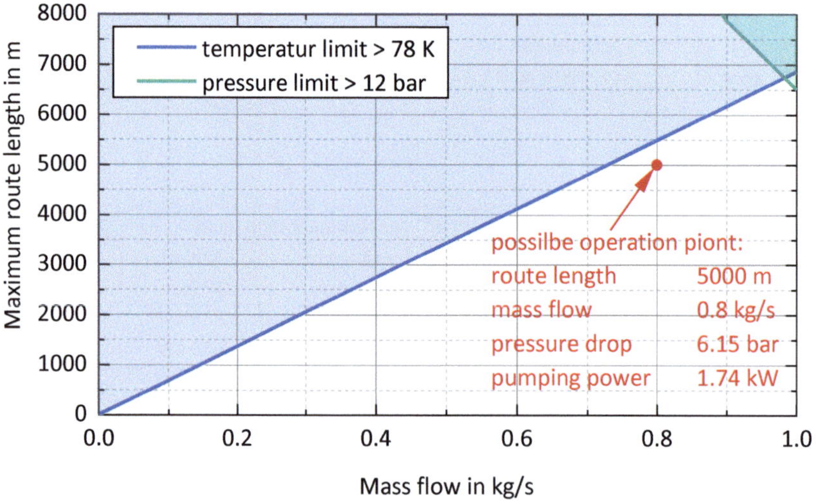

Figure 3.12: Representation of the maximum route length as a function of the mass flow with two-sided cooling for the nominal width DN 50 of the inner tube.

In Table 3.15 the maximum route lengths are listed for halved cryostat transmission losses. A reduction of cryostat losses could be achieved with rigid cable cryostats. As an example the cryostat manufacturer CryoTherm gives a loss value of 1.3 W/m for a rigid cryostat at DN 100 [CRY15]. The following figure serves to illustrate the development potential and to provide an incentive for possible further investigations. For the current design, the losses of a flexible cryostat are still assumed.

Table 3.15: Maximum route lengths for cooling on one and both sides with reduced cryostat transmission losses depending on the nominal diameter of the inner tube (hollow core conductor concept) and the mass flow

Nominal diameter of inner pipe	Mass flow	Maximum route length in *m*	
		One-sided cooling	Double-sided cooling
DN 32	$\dot{m}$ = 0,5 kg/s	2415 → 4130	3630 → 6200
	$\dot{m}$ = 1,0 kg/s	4830 → 8265	7250 → 12400
DN 40	$\dot{m}$ = 0,5 kg/s	2470 → 4280	3710 → 6420
	$\dot{m}$ = 1,0 kg/s	4940 → 8560	7410 → 12850
DN 50	$\dot{m}$ = 0,5 kg/s	2380 → 4235	3570 → 6350
	$\dot{m}$ = 1,0 kg/s	4760 → 8470	7140 → 12710

Halving cryostat transmission losses from Table 3.10 results in an average 42% increase in route length. An almost identical increase in the results can also be observed for the hollow core conductor concept (Annex, Table A.1).

3.1.4 Superconducting requirement

Decisive for the superconducting requirement of the conductor layer are the route length, the number of SC tapes, the angle of twist, and the critical current of the twisted SC tapes. the schematic diagram in Figure 3.13 is used to illustrate the relationship between SC tape arrangement and lay angle of the individual SC tape.

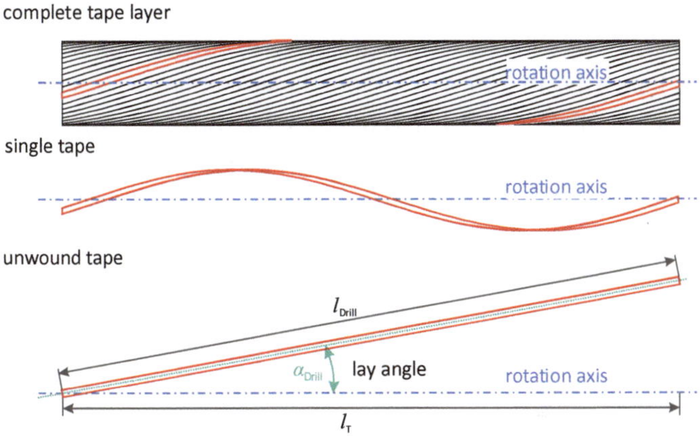

Figure 3.13: Schematic representation of the relationship between SC tape arrangement and lay of the single SC tape.

Figure 3.13 shows that the unwound length of the SC tape l_{Drill} is greater than the route length l_{T}. The geometric relation between route length and lay angle α_{Drill} is shown in equation 3.31. The assumed angle of shock in this study is 15°.

$$l_{\text{Drill}} = \frac{l_{\text{T}}}{cos(\alpha_{\text{Drill}})} \tag{3.31}$$

3.1.4.1 SC tape requirement: electrical phase

Based on the developed SC tape length and the number of SC tapes per HTS layer, the superconductor requirement can be determined according to equation 3.32.

$$l_{\mathrm{L,ges}} = \underbrace{3}_{\text{3 phasiges System}} \cdot \sum_{i=1}^{N_{\mathrm{LS}}} N_{\mathrm{L,i}} \cdot l_{\mathrm{Drill}} \tag{3.32}$$

$l_{\mathrm{L,ges}}$ Superconductor requirement per three-phase system (conductor layer) (km)
N_{LS} Number of conductor layers
$N_{\mathrm{L,i}}$ Number of SC tapes per layer
l_{Drill} Developed SC tape length (km)

The number of SC tapes per layer is determined by the inner tube diameter or the inner tube circumference. The current carrying capacity and the number of layers depend on the critical current I_{C} of the SC tapes. In the dimensioning of this conceptional design, a critical electric, considering the temperature and the magnetic field, of 150 A is assumed for a SC tape with the width of 4 mm.

If this value is increased to 200 A, a SC tape layer can be saved with an inner pipe nominal diameter of DN 40 and DN 50 (equation 3.8). However, the resulting increase in the load factor F increases the AC losses in the SC tapes. This results in a reduction of the maximum route length, as shown in Figure A.7 and Figure A.8 in the Annex. By adjusting the critical current, it is thus possible to set an optimum between the SC tape requirement and maximum route length. From a monetary point of view, this means an optimisation between investment and operating costs. However, an optimization of investment and operating costs is not part of this study. In Table 3.16 the superconductor requirement of the conductor layer is listed as a function of the critical current and the inner tube.

Table 3.16: Overview of the superconductor requirement (three-phase, one system) of the conductor layer as a function of the critical current and of the inner tube (hollow core conductor concept)

Nominal diameter Inner tube	geo. Number of SC tapes Layer 1/layer 2	Superconducting requirement for 1 km route length (three-phase)	geo. Number of SC tapes Layer 1/layer 2	Superconducting requirement for 1 km route length (three-phase)
	$I_{\mathrm{C}} = 150$ A		$I_{\mathrm{C}} = 200$ A	
DN 32	31/31	192.6 km	31/31 a)	192.6 km a)
DN 40	36/37	226.7 km	36/0	111.8 km
DN 50	44/44	273.3 km	44/0	136.7 km

Note: a) Equation 3.8 results in a minimum number of electrical SC tapes (Ic = 200 A) of 32, so that the variant DN 32 must still be designed with two HTS layers. The lay angle θ is assumed to be 15°.

If the nominal diameter DN 50 of the inner tube (hollow core conductor concept) is selected, a conductor layer with 44 SC tapes can be saved by increasing the critical current.

Table 3.7 shows that the number of SC tapes required is sufficient with just one SC tape layer. This means that 136.7 km of SC tapes can be saved per 1 km of route length.

3.1.4.2 SC tape requirement: magnetic shield layer

If the external magnetic field of the single-core cable is to be suppressed, at least one additional HTS shield layer is added. The calculation of the superconductor requirement for the shielding layer is carried out according to equation 3.33.

$$l_{\mathrm{S,ges}} = \underbrace{3}_{\text{3 phasiges System}} \cdot \sum_{i=1}^{N_{\mathrm{SS}}} N_{\mathrm{S,i}} \cdot l_{\mathrm{Drill}} \tag{3.33}$$

$l_{\mathrm{S,ges}}$	Superconductor requirement per three-phase system (shield layer)	(km)
N_{SS}	Number of shield layers	
N_{S}	Number of SC tapes in the shield layer	

The shielding layer is located above the electrical insulation, as shown in Figure 3.1. This diameter or circumference is relatively large for extra-high voltage cables. A high number of SC tapes is required to ensure that the shield layers are equipped as completely as possible. In Table 3.17 the superconductor requirement of the shielding layer is listed as a function of the inner tube.

Table 3.17: Overview of the superconductor requirement (three-phase, one system) of the shielding layer depending on the inner tube (hollow core conductor concept)

Nominal diameter Inner tube	geo. Number of SC tapes	Superconducting requirement for 1 km route length (three-phase)
DN 32	71	220.5 km
DN 40	74	229.8 km
DN 50	81	251.6 km

The superconductor requirement of a three-phase cable system with a nominal diameter of DN 50 (hollow core conductor concept) is 388.3 km per km of line length. The magnetic shielding layer requirement accounts for approximately two thirds of the total requirement.

3.2 Cable properties

3.2.1 Electrical properties (RLC parameters)

This study defines electrical properties as resistance, inductance and capacitance quantities per unit length. Analytical equations exist for the calculations of these quantities per unit length of the single conductor concept. The resistance per unit length is very small due to the superconducting state and can be assumed with $R' < 0.0001\ \Omega/\text{km}$ [HKS03]. The inductance per unit length is calculated according to equation 3.34.

$$L' = \frac{\mu_0}{4\pi} tan(\theta_{BL})^2 + \frac{\mu_0}{2\pi} \cdot log\left(\frac{r_{\mathrm{S,in}}}{r_{\mathrm{L,au}}}\right) + \frac{\mu_0}{2\pi\left(r_{\mathrm{L,au}}^2 - r_{\mathrm{L,in}}^2\right)}\left[\frac{r_{\mathrm{L,au}}^4 - r_{\mathrm{L,in}}^4}{4} + r_{\mathrm{L,in}}^4 log\left(\frac{r_{\mathrm{L,au}}}{r_{\mathrm{L,in}}}\right) - r_{\mathrm{L,in}}^2\left(r_{\mathrm{L,au}}^2 - r_{\mathrm{L,in}}^2\right)\right] \tag{3.34}$$

L'	Inductance per unit length	(H/m)
μ_0	Magnetic field constants ($4\pi \cdot 10^{-7}$)	(Vs/Am)
θ_{BL}	Lay angle of the SC tapes(15°)	(°)
$r_{\mathrm{S,in}}$	Inner radius of the shield	(m)
$r_{\mathrm{L,in}}$	Inner radius of the conductor	(m)
$r_{\mathrm{L,au}}$	Outer radius of the conductor	(m)

The first term of the equation describes the inductance based on the lay angle, the third term considers the finite thickness of the conductor layer. These two terms have no significant influence (approx. 5 %) on the total inductance. The main influencing factor is the magnetic field between conductor and shield, calculated by the second term of equation 3.34. The capacitance per unit length is calculated according to equation 3.35.

$$C' = \frac{2\pi \cdot \varepsilon_0 \cdot \varepsilon_{\mathrm{TY}}}{ln\left(\frac{r_{\mathrm{EI,au}}}{r_{\mathrm{EI,in}}}\right)} \tag{3.35}$$

C'	Capacitance per unit length	(F/m)
ε_0	Electrical field constant	(As/Vm)
$\varepsilon_{\mathrm{TY}}$	Relative permittivity of the insulation (Tyvek™)	(1.73 [HNM14])

The capacitive charging current (operation with open line end) of a cable can be calculated with equation 3.36 [OO11].

$$I'_{\text{Lade}} = \omega \cdot \frac{U_{\text{N}}}{\sqrt{3}} \cdot C' \tag{3.36}$$

I'_{Lade}	Capacitive charging current	(A/m)
ω	Angular frequency	(1/s)
U_n	Nominal voltage	(V)
l	Length of the cable	(m)

The charging power of a cable can be calculated with equation 3.37 [0011].

$$Q'_{\text{Lade}} = \sqrt{3} \cdot U_{\text{N}} \cdot I'_{\text{Lade}} = \omega \cdot U_{\text{N}}^2 \cdot C' \tag{3.37}$$

Q'_{Lade}	Capacitive charging current	(VA/m)

Table 3.18 provides an overview of the inductance and capacitances per unit length of the single-conductor cable as well as the charging capacity in dependency of the nominal diameter.

Table 3.18: Provides an overview of the inductance and capacitance per unit length of the single-conductor cable as well as the charging capacity based on the nominal diameter.

			Nominal diameter DN 32	Nominal diameter DN 40	Nominal diameter DN 50
Hollow core conductor concept	Inductance per unit length L'	(mH/km)	0.169	0.146	0.129
	Capacitance per unit length C'	(nF/km)	119.5	140.2	158.8
	Charging power Q'_{Lade}	(MVar/km)	5.4	6.4	7.2
Corrugated tube concept	Inductance per unit length L'	(mH/km)	0.201	0.163	0.136
	Capacitance per unit length C'	(nF/km)	100.2	124.3	150.5
	Charging power Q'_{Lade}	(MVar/km)	4.5	5.6	6.9

Both carrier concepts, inner tube and hollow core conductor, have similar inductance and capacitance quantities per unit length. While XLPE cables have a capacitance per unit length of about 240 nF/km, their capacitance per unit length is about two-thirds of that of conventional cables. In relation to the entire transmission system, this results in a reduction of more than 65 %. The charging capacity for a 5 km long partial cabling thus amounts to about 36 MVar. The inductance per unit length of a superconducting cable is approx. 20 % compared to that of a conventional cable [HKS03].

3.2.1.1 Comparison of operating characteristics - line constants

Table 3.19 shows the line constants of the various transmission systems in comparative form.

Table 3.19: Overview of the line constants

		Overhead line	Underground cable	Super conducting cable	Gas insulated cables
Inductance per unit length L'	(mH/km)	0.80	0.48	0.13	0.2
Capacitance per unit length C'	(nF/km)	13	230	158	55-70
Resistance per unit length R'	(mΩ/km)	36	7.2	>1	

3.2.1.2 Comparison of operating characteristics - reactive power requirement, charging current, transmittable power

The transmittable power of the considered line types on a route length of < 10 km is not influenced by the charging power and the voltage drop. The Ferranti effect can be neglected with < 0.05 % over a route length of < 10 km.

3.2.2 Calculation of the losses

Compared to conventional cables, superconducting cables are extremely efficient, especially at high current loads. Nevertheless, various losses occur during operation, which can be classified as follows:

- Alternating current losses in superconductor material
- Dielectric losses
- Thermal losses of the line cryostat
- Losses in terminations and power supplies

3.2.2.1 Alternating current losses in superconductor material

The theoretical background to alternating current losses in superconducting cables has already been discussed in 3.1.2.6.

The AC power loss depends on the temperature profile of the cable for a specified route length, mass flow and current load. The left part of the Figure 3.14 shows the temperature profile in the HTS SC tape layer at a nitrogen mass flow of 0.9 kg/s with one sided cooling and a route length of 3.2 km with a cable with a nominal diameter of the inner pipe of DN 50 and a single HTS SC tape layer at full rated current. The corresponding power loss as a function of the route length is shown in the right part of the Figure 3.14.

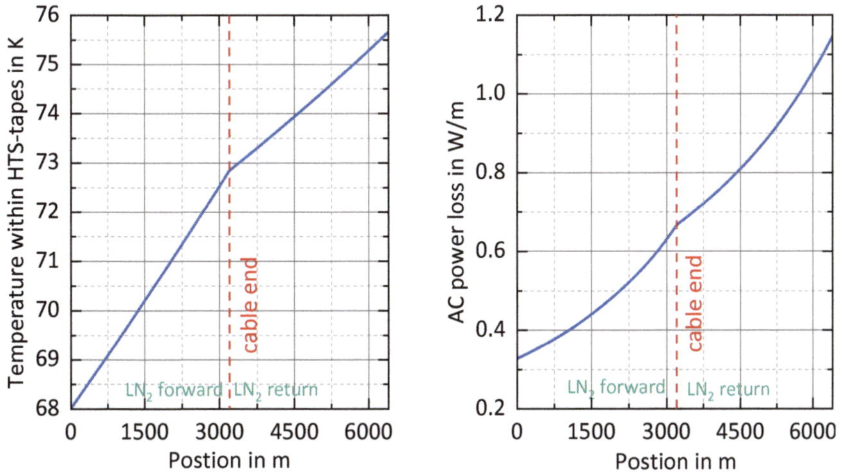

Figure 3.14: Temperature curve in the HTS SC tape layer as a function of the route length (left); AC power loss per phase as a function of the route length (right).

Table 3.20 provides an overview of the integrated AC power loss at various current loads over the entire route length. The total power loss includes the power loss of the three electrical phases L1, L2 and L3 as well as the corresponding losses in the various shields. This results in six loss components.

Table 3.20: Overview of the total AC power losses (three-phase) for a system with different current loads and a route length of 3.2 km

	Nominal diameter of former	Number of SC tapes per layer (S1/S2/SS)	Power loss	Power loss	Power loss
			$0{,}1 \cdot I_N$ in W	$0{,}5 \cdot I_N$ in W	$1 \cdot I_N$ in W
Hollow core conductor concept	50	44/0/81	< 1	331	6594

The number of SC tapes in the cable system decreases with decreasing nominal inner pipe diameter; at the same time, alternating current losses increase sharply. This results in a greater power loss and thus a reduction in the distance between cooling stations. On the other hand, investment costs are reduced by reducing the need for SC tapes.

3.2.2.2 Dielectric line losses

The calculation of dielectric losses P_{V0D} of a cable has already been described in section 3.1.2.6, heat source - dielectric losses of electrical insulation. In Table 3.21 dielectric losses are shown.

Table 3.21: Overview of the dielectric losses per cable

Relative permittivity	Insulation thickness	Inner diameter	Capacitance per unit length	Loss factor	Dielectric power loss
	mm	mm	pF/m	$tan\delta$	W/m per cable
1.73	24.25	58.2	158.8	$1{,}3 \cdot 10^{-4}$	0.31

3.2.2.3 Cryostat transmission losses

Thermal losses occur in the line cryostats $P_{\mathrm{V,0,LK}}$, i.e. depending on the thermal insulation quality. The environment of the cryostat is always warmer than that of the cable system, resulting in a permanent heat flow in the direction of the cold part. This results in load-independent losses that must be constantly compensated by the nitrogen cooling system. The heat transfer can be divided into heat radiation, heat conduction and convection. The heat conduction in the line cryostats is greatly reduced by small contact surfaces. Convection is suppressed by a long-term stable vacuum as an insulating layer between the cryostat envelopes.

The remaining thermal losses are therefore thermal radiation losses. These are dependent on the thermal conductivity λ of the super insulation, whose values lie between 0.0025 W/(m K) and 0.005 W/(m·K). The thickness of the insulation is typically 7-10 mm. The data from Table 3.10 can be used to calculate the losses of line cryostats. Typical values are 2 W/m [CRY12].

3.2.2.4 Power supply losses

Thermal and electrical losses occur in the power supply systems. The thermal losses occur at the transition between the low temperature range and the ambient temperature of the termination cryostat. Electrical losses in the power supply lines are a result of the resistive metallic supply lines and depend, among other things, on the type of superconductor cable. The minimum heat input of a bath cooled power supply is approx. 42 W/kA per pole at a lower temperature level of 77 K [HCB93].

Since no detailed design of a power supply can be carried out in this study, the total power supply losses are assumed at 45 W/kA for the following calculations. Experience has shown that thermal losses at rated current account for 37 % of losses. 63% are load-dependent electrical losses. Equations 3.38 and 3.39 show the calculation of the power supply losses for each termination.

$$P_{\mathrm{V,SZ}} = \left(P_{\mathrm{V,th,SZ}} + P_{\mathrm{V,el,SZ}}\right) \tag{3.38}$$

$$P_{\mathrm{V,SZ}} = 45\frac{\mathrm{W}}{\mathrm{kA}} \cdot I_{\mathrm{r}}\left(37\% + 63\% \cdot \left(\frac{I}{I_{\mathrm{r}}}\right)^3\right) \tag{3.39}$$

$P_{\mathrm{V,SZ}}$	Critical current as a function of temperature	(W)
$P_{\mathrm{V,th,SZ}}$	Thermal power supply losses	(W)
$P_{\mathrm{V,el,SZ}}$	Electrical power supply losses	(W)
I_{r}	Rated current	(A)
I	Nominal current	(A)

Table 3.22 summarises the power supply losses.

Table 3.22: Overview of power supply losses per termination at different load factors

Nominal current	Symbol	Power loss	Power loss	Power loss
A		$0{,}1 \cdot I_N$ in W	$0{,}5 \cdot I_N$ in W	$1 \cdot I_N$ in W
3600	$\boldsymbol{P_{V,SZ}}$	61.0	87.1	168.5
	$P_{V,th,SZ}$	59.9	59.9	59.9
	$P_{V,el,SZ}$	1.1	27.1	108.5

3.2.2.5 Losses of termination cryostats

The thermal losses of the termination cryostats $P_{V,0,EK}$ are load-independent and are assumed to be 20 W each (empirical value).

3.2.2.6 Cryocooler operation

The electrical connected loads of the cryocoolers are treated as losses. In principle, the required connected load of a cryocooler can be estimated from the efficiency η of the cooling process according to equation 3.40. For lines with cold dielectric, the dielectric losses $P_{V,0,D}$ must be taken into account, since the resistive component in the dielectric loss heats the line. The efficiency depends on the cooling system. This study assumes an average efficiency of approx. 6.3 % (68-300 K) [SC12].

$$P_{V,KK} = \frac{\left(P_{V,SL} + P_{V,0,LK} + P_{V,SZ} + P_{V,0,EK}\right)}{\eta} \tag{3.40}$$

$P_{V,KK}$	Connected load of the cryocooler	(W)
$P_{V,SL}$	Alternating current losses	(W)
$P_{V,0,LK}$	Cryostat transmission losses	(W)
$P_{V,SZ}$	Power supply losses	(W)
$P_{V,0,EK}$	Cryostat termination losses	(W)
η	Efficiency of the cryocooler	

For the subsequent annual energy loss calculation, the individual loss shares are determined separately.

3.2.2.7 Overview of loss calculation

In Table 3.23 you will find an overview of the total losses of the two insulated cables projected for a current intensity of 3.6 kA and a route length of 3200 m. The results apply to the hollow core conductor concept with the former nominal diameter DN 50 and the inner cryostatic tube DN 125.

Table 3.23: Overview of the losses for two systems with a rated current of 3.6 kA and a route length of 3200 m (hollow core conductor concept)

Loss component	Symbol	Power loss	Power loss	Power loss
		$0,1 \cdot I_N$	$0,5 \cdot I_N$	$1 \cdot I_N$
Cooling capacity	$\boldsymbol{P_V}$	45329 W	46303 W	59806 W
Alternating current losses	$P_{V,SL}$	1.0 W	662 W	13188 W
Dielectric losses	$P_{V,0,D}$	5956 W	5956 W	5956 W
Cryostat transmission losses	$P_{V,0,LK}$	38400 W	38400 W	38400 W
Power supply losses	$P_{V,SZ}$	732 W	1045 W	2022 W
Cryostat termination losses	$P_{V,0,EK}$	240 W	240 W	240 W
Ambient temperature losses	$\boldsymbol{P_{V,g}}$	719.5 kW	735.0 kW	949.3 kW

3.2.2.8 Reduction of the losses of line cryostats

Table 3.23 shows that most of the losses are caused by the line cryostats. Table 3.24 shows the loss components under the assumption that the cryostat transmission losses have been halved. Such a reduction in cryostat losses could be achieved with greatly improved rigid line cryostats. CryoTherm, the cryostat manufacturer, specifies a loss value of 1.3 W/m for its rigid cryostats at DN 100 [CRY15]. According to Cryoflex, a flexible system has a loss value of 2.2 W/m at DN 100. However, the loss data of CryoTherm refers only to the line cryostats - couplings, sleeves and connecting parts are not included.

Table 3.24: Overview of the losses for two systems with a rated current of 3.6 kA and a route length of 3200 m under the assumption of halved cryostat transmission losses

Loss component	Symbol	Power loss	Power loss	Power loss
		$0,1 \cdot I_N$	$0,5 \cdot I_N$	$1 \cdot I_N$
Cooling capacity	$\boldsymbol{P_V}$	26129 W	27103 W	40606 W
Alternating current losses	$P_{V,SL}$	1.0 W	662 W	13188 W
Dielectric losses	$P_{V,0,D}$	5956 W	5956 W	5956 W
Cryostat transmission losses	$P_{V,0,LK}$	19200 W	19200 W	19200 W
Power supply losses	$P_{V,SZ}$	732 W	1045 W	2022 W
Cryostat termination losses	$P_{V,0,EK}$	240 W	240 W	240 W
Ambient temperature losses	$\boldsymbol{P_{V,g}}$	414.7 kW	430.2 kW	644.5 kW

3.2.3 Mechanical properties

The bending radius of superconducting cables is assumed to be the same as the bending radius of XLPE cables. The rule of thumb is 25 times the outer diameter. This results in a bending radius of 4.3 m with a cable diameter of 172 mm.

3.3 Short-circuit calculation

IEC 60909 standard was used for the following short-circuit calculations. The calculation of the short-circuit behaviour is taken over from [Str11]. An adiabatic state is assumed for the calculation, i.e. there is no heat exchange with the nitrogen.

In Figure 3.15 the short-circuit currents of the cable are shown. On the one hand, the short-circuit current in the network without connected cable is considered; on the other hand, the short-circuit current including HTS cable is calculated. The nominal diameter is DN 50, the cable length 5 km. A three-pole short-circuit with an RMS value of 20 kA is assumed.

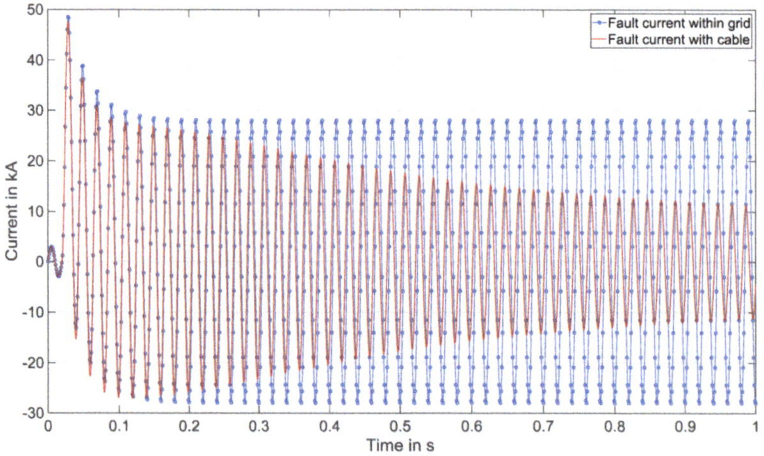

Figure 3.15: Short-circuit currents in the inner tube concept in the grid and in the cable with a three-pole short-circuit current of 20 kA. The nominal diameter of the inner tube is DN 50, the cable length is 5 km.

A current-limiting behaviour in the superconducting cable can be detected. This behaviour is caused by the so-called quench of the SC tapes. This means that they lose their superconducting property. The current then flows through the other layers of the SC tape.

Figure 3.16 shows the short-circuit currents of the cable with hollow core conductor in the network and in the cable. The nominal diameter is DN 50, the cable length 5 km. A three-pole short-circuit with an RMS value of 20 kA is assumed.

With the hollow core conductor concept, which is equipped with a 300 mm^3 copper stabilisation, no current-limiting behaviour can be detected. The additional thermal and electrical stabilization prevents the loss of the superconducting state.

As worst case estimation a starting temperature at the cable end of 76 K is assumed. The table summarizes the SC tape temperatures at the end of the cable after a specified short-circuit duration.

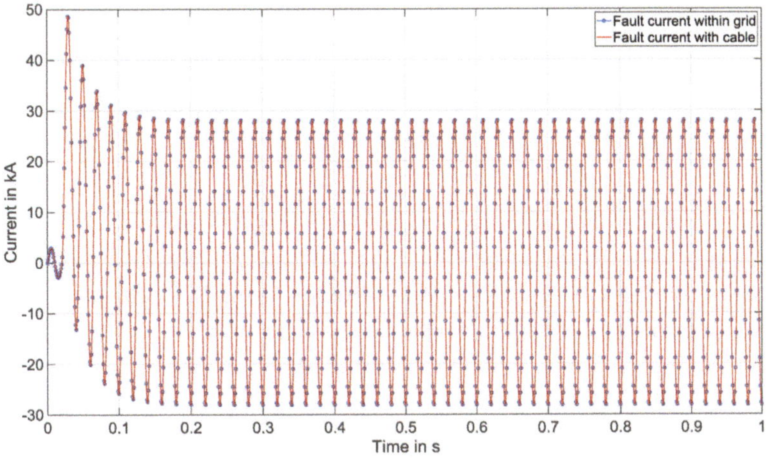

Figure 3.16: Short-circuit currents in the hollow core conductor concept in the mains and in the cable with a three-pole short-circuit current of 20 kA. The nominal diameter of the inner tube is DN 50, the cable length is 5 km.

Table 3.25: Overview of the SC tape temperatures after defined short circuit duration of a three-pole short circuit

	Short-circuit current (RMS value)	Short-circuit data	SC tape temperature at cable end
	kA	ms	K
Corrugated tube concept	20	150	187
		300	432 a)
		1000	1639 a)
	63	150	773 a)
		300	1158 a)
		1000	2236 a)
Hollow core conductor concept	20	150	76
		300	76
		1000	77
	63	150	78
		300	81
		1000	94

Note: The assumed approximation functions for the temperature-dependent material properties are valid for temperatures up to 300 K. Deviations from these values occur.

The results show the necessity of additional copper stabilisation. This prevents irreversible damage to the HTS SC tapes at temperatures of > 175°C [Bay16[3]]. In addition, the formation of gaseous nitrogen can be avoided by maintaining a corresponding minimum operating pressure (> 3 bar).

3.4 Conceptual Design

The final cable design meets all network requirements for the cable system. For further loss calculations, a final cable design is selected based on the project parameters. Table 3.26 provides an overview of the characteristics of the final cable design.

[3] https://doi.org/10.1088/0022-3727/3/4/308

Table 3.26: Overview of the characteristics of the final cable design

Rated current per system	3600 A
Rated voltage	420 kV
Short-circuit data	63 kA for 300 ms
Route length	3200 m
Cable concept	Single conductor with two systems (with copper stabilization)
Cooling concept:	one-sided cooling concept
Maximum pressure loss	6.3 bar
Maximum global temperature at rated current	77.2 K
Maximum global temperature in the event of a short circuit	100.2 K
Pump capacity	1.0 kW
Mass flow	0.9 kg/s
Capacitance per unit length	158.8 nF/km
Inductance per unit length	0.129 mH/km
Charging capacity (per system)	23 MVar
Total losses at rated current and ambient temperature	644.5 kW
HTS SC tape requirement: one system per route kilometre	388.3 km
HTS SC tape requirement: a system (@ 3.2 km]	1242.6 km
HTS SC tape requirement: two systems	2485.2 km

Figure 3.17 shows the CAD model of the selected final cable design. Table 3.27 indicates the geometric cable build-up of the conceptional design.

Figure 3.17: Conceptual design of a superconducting extra-high voltage cable.

Table 3.27: Cable build-up of the conceptional design

Diameter	Symbol	Size
Carrier (i)	$d_{T,i}$	50.0 mm
Carrier (a)	$d_{T,a}$	54.0 mm
Hollow core conductor (a)	$d_{HL,a}$	57.4 mm
SC tape layer (a)	$d_{L,a}$	57.8 mm
Number of SC tapes	N_L	44 parts
Electrical isolation (a)	$d_{EI,a}$	106.3 mm
Shield layer (a)	$d_{SS,a}$	106.7 mm
Number of shielding conductors	N_S	81 parts
Neutral conductor (a)	$d_{NL,a}$	110.7 mm
Inner cryostat tube (i)	$d_{IK,i}$	125.0 mm
Inner cryostat tube (a)	$d_{IK,a}$	130.2 mm
Outer cryostat tube (i)	$d_{AK,i}$	150.0 mm
Outer cryostat tube (a)	$d_{AK,a}$	162.0 mm
Jacket (a)	$d_{M,a}$	172.8 mm

The outer diameter of the cable is approx. 173 mm; this allows the cable to be pulled into a protection tube with a nominal diameter of 200 or more.

3.5 Cooling system

Several cooling systems for the required temperature range are available on the market. When the main requirement is to keep maintenance costs as low as possible, Stirling coolers and Turbo-Brayton coolers appear to be the most attractive variants.

The Stirling cooler provides cooling capacity up to 3 kW @ 65K. A Turbo-Brayton cooler (also known as a Brayton cooler) is used for the two-digit kilowatt range. The high required cooling capacity of more than 40 kW would require a high number of units of more than ten if implemented with Stirling type cooler, so that this study accepts the Brayton cooler.

A Brayton cooler consists of two active parts, a compressor and a turbine, as well as two passive heat exchangers. It is based on the principle of the reversal of a Brayton machine, as it is used in every gas or coal-fired power plant to generate energy. Instead of using hot combustion gases, however, the cooler usually works with a neon gas in a closed system. The compressor compresses the gas to about 10 bar and heats it up. The gas is then cooled with cooling water in an aftercooler before the temperature is lowered to below 100 K in a countercurrent heat exchanger. The gas is then expanded in a turbine to about 6 bar and cooled down to the desired temperature level. The existing cooling capacity from the capacity flow $(\dot{m} \cdot c_{\mathrm{p}})_{\mathrm{Neon}}$ can now be used with a countercurrent heat exchanger to undercool the nitrogen. Figure 3.18 shows the basic structure.

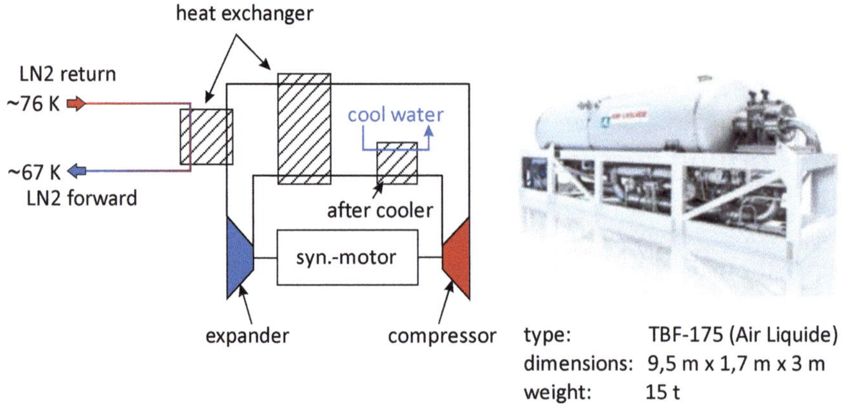

Figure 3.18: Schematic design and dimensions of a Brayton cooler type TFB-175 (Air Liquide).

The advantages of the Brayton cooler shown here result from the completely oil and contact-free operation with high-speed synchronous motors. This increases the maintenance intervals, which reduces operating costs. Furthermore, the cooler does not require a gearbox, which would lead to additional wear and power losses. Compressed air, which many older models use for compression, can also be dispensed with. Further technical parameters are listed in Table 3.28.

Table 3.28: Characteristics of some coolers of different manufacturers and cooling methods

Label	Air Liquide TFB-175	Taiyo Nippon Sanso Corporation[4]	Striling Cryogenics SPC-4 Cryogenerator
Cooling capacity	15 kW @ 70 K	10 kW @ 70 K	4 kW @ 70 K
Number of devices for the overall system	4	5	13
Efficiency	~ 8.4 %	~ 8.0 %	~ 6.4 %
Maintenance interval	5 years	-	>6000 h
Max. mass flow	-	0.7 kg/s	-
Dimensions	9.5 x 1.7 x 3 m	7.5 x 2.3 x 2.3 m	1.7 x 0.8 x 1.2 m
Mass	15 t		1.2 t

Note: The cooling capacity is based on the power losses Table 3.24.

[4] https://doi.org/10.1063/1.4860847

The TBF type Brayton coolers are partial load capable. In Figure 3.19 the cooling capacities are shown as a function of temperature for types TBF-80, TBF-175 and TBF-350.

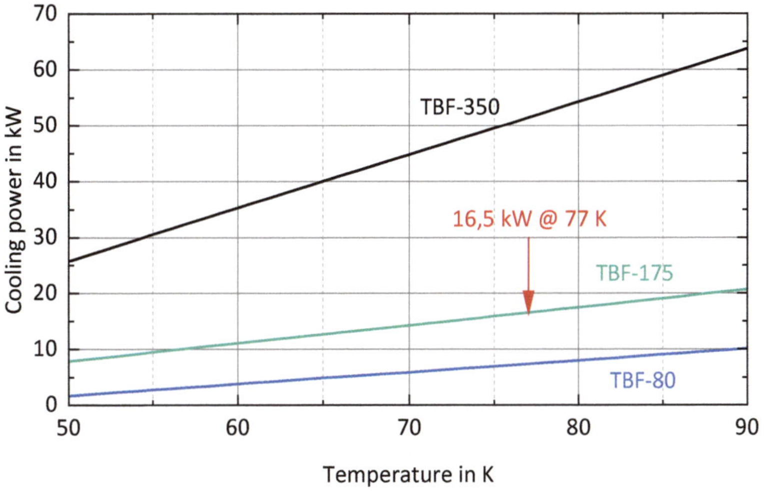

Figure 3.19: Representation of the cooling capacity of Brayton coolers type TBF-80/175/350 as a function of temperature [AAL18].

For redundancy reasons, the chiller is designed for the cable system with four TBF-175 coolers. In the event of maintenance work or a technical fault, the transmission line can continue to operate without interruption in accordance with the (n - 1) criterion.

The total connected load of the chiller of approx. 650 kW can be obtained via the internal transmission grid of the transmission grid operators (internal supply from the point of view of the transmission grid operator) or via a local distribution grid (external supply). This distinction has a decisive influence on the economic efficiency of the system (see section 4.1.3).

3.5.1 Schematic diagram of system

Figure 3.20 shows a simplified schematics - without additional safety technology - for cooling the conceptual cable. The cooling system consists of four Turbo Brayton coolers that undercool the liquid nitrogen. The necessary flow rate is adjusted by a cryogenic liquid nitrogen pump, which can also work with gaseous nitrogen. This technical property can be used for the cooling process with gaseous nitrogen. A further pump is installed in the circuit for redundancy. The reservoir, which is filled with liquid nitrogen from the LN2 tank, performs several functions during the cooling of the cable, which are described in more detail below.

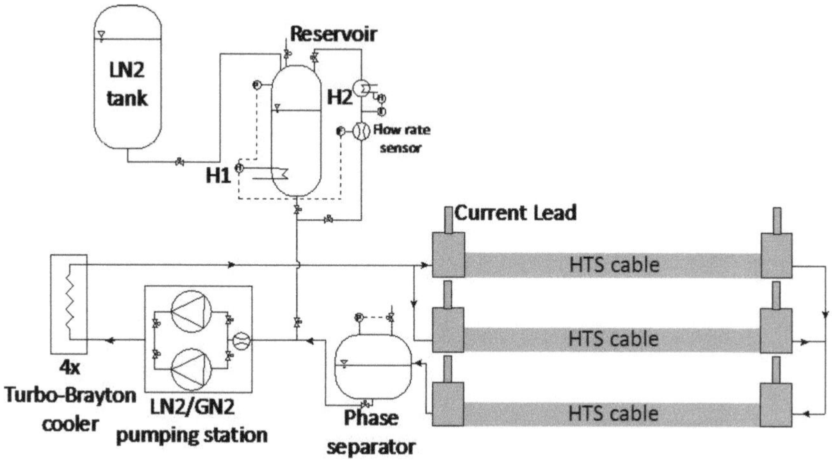

Figure 3.20: Schematic setup of the cooling of the conceptual cable.

3.5.2 Cooling of the cable

The superconducting cable must be cooled from the ambient temperature to the supercooled nitrogen temperature before switching on. To avoid excessive thermal stresses in the axial and radial directions during cooling, it is proposed in [YMA06[5]], [CKK12[6]] and [MSH05[7]] to divide the process into three sub-steps:

1. Gas-cooled cooling at different temperatures, mass flows and waiting times (precooling process). The liquid nitrogen is evaporated with a heater H1 in a reservoir tank and heated to the desired temperature with an additional heating element H2. The mass flow of the evaporating nitrogen is controlled by an associated flow sensor (see Figure 3.21).

[5] https://doi.org/10.1063/1.2202486
[6] https://doi.org/10.1109/TASC.2011.2178376 [7] https://doi.org/10.1109/TASC.2005.849289

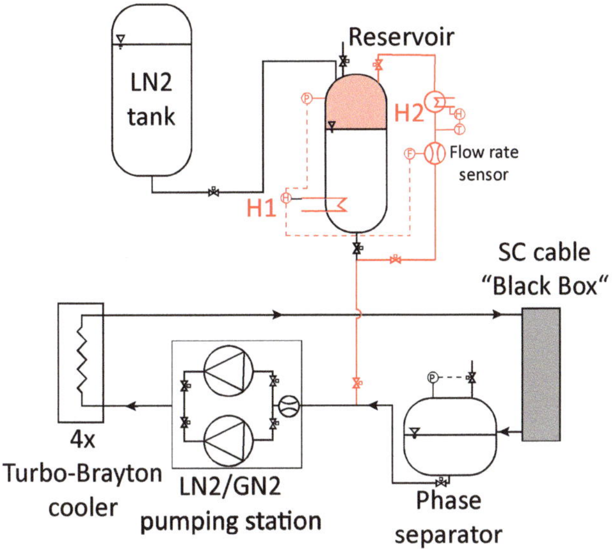

Figure 3.21: Lines and fittings (red) for cooling in sub-step 1.

Filling the cable with liquid nitrogen from the reservoir in saturation condition at approx. 2-3 bar (see Figure 3.22).

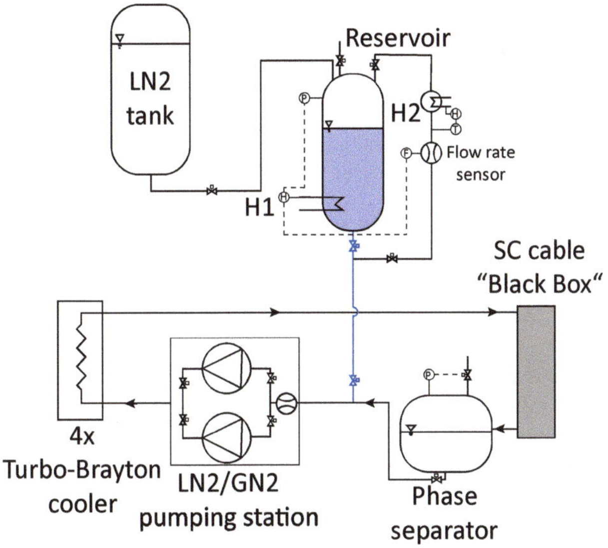

Figure 3.22: Lines and fittings (blue) for cooling in sub-step 2.

Switching on the chiller and pump to cool the cable down to the temperature of the supercooled nitrogen (see Figure 3.23).

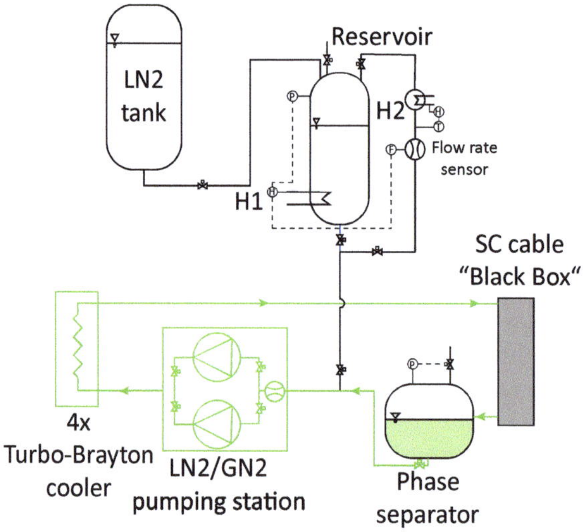

Figure 3.23: Lines and fittings (green) for cooling in sub-step 3.

The conservative values assumed in Table 3.29 can be used to estimate the temperature ranges, mass flows and waiting times. The cooling time of superconducting cables varies with the length, but is in the area of about 150 hours.

Table 3.29: Estimation of the necessary times and mass flows of nitrogen for different temperatures in the respective sub-steps according to the method in [CKK12].

Fractional pitch	1				2	3
Temperature (K)	200	150	100	80	80	65
Time (hours)	16	20	20	70*	> 10	> 10
Mass flow (kg/s)	$0.01 < \dot{m} < 0.02$				Periodic inflow	Nominal operation

* The long waiting time is required until the HTS cable has cooled down to the temperature of 80 K at the outlet.

Before sub-step 1, it is recommended to flush the cable with heated gaseous nitrogen at temperatures above 60 °C. The accumulated moisture in the cable is removed and other components of the air are flushed out. This also minimizes the risk of oxygen condensing in the cable during cooling.

3.5.3 Warm-up times in the event of failure of the chiller

If the chiller and the pump fail as a result of a supply interruption, the liquid nitrogen flow comes to a standstill. The absorbed heat power of nitrogen can no longer be transferred to the chiller and, due to heat radiation, the nitrogen initially heats up to the boiling temperature at the respective pressure; it is then completely evaporated. The following calculations are used to estimate the warm-up time of the cable in the event of system failure.

3.5.3.1 Heating of the liquid nitrogen up to the bubble-point curve

If the pump fails, the pressure in the system adjusts to the low pressure level of 3 bar. For the calculation of the times a temperature of 78 K is assumed in the complete cable. In fact, shortly after the cooling failure, there is still an axial temperature profile as in Figure 3.14. This underestimates the calculated times for heating the nitrogen.

To heat the nitrogen to the boiling point $T_S \approx$ of 88 K at a pressure of 3 bar, the energy

$$Q_1 = m_{LN2} \cdot c_{p,LN2} \cdot (T_S - 78\,K) \approx 100\,MJ \tag{3.41}$$

is required. The mass of liquid nitrogen in the cable is:

$$m_{LN2} = \frac{\pi \cdot {d_{T,i}}^2}{4} \cdot l \cdot \rho_{LN2} \approx 5050\,kg \tag{3.42}$$

l Stands for the length of the cable in (m), ρ_{LN2} for the density in (kg/s) and $d_{T,i}$ for the inner diameter of the carrier tube in (m).

The warm-up time to T_S can be calculated with the heat input distributed by radiation over the entire cable and amounts to:

$$t_1 = \frac{Q_1}{q_{rad} \cdot l} \approx 4\text{ Stunden.} \tag{3.43}$$

When heating up to the boiling temperature, T_S it should be noted that the density of the liquid nitrogen changes and thus the volume occupied by the liquid increases. The change in volume ΔV in (L) is between 78 K and 88 K:

$$\Delta V = \frac{m_{LN2}}{\rho_{LN2}(88K, 3bar)} - \frac{m_{LN2}}{\rho_{LN2}(78K, 3bar)} \approx 400\,L. \tag{3.46}$$

The relatively large expansion to 400 litres requires an additional buffer tank to compensate for the thermal expansion of the nitrogen.

If an emergency power supply is available for the LN2 pumps, they can be operated further in order to increase the pressure level of the liquid nitrogen and thus the distance to the bubble-point curve. In addition, the convective heat transport prevents the accumulation of heat in the cable. This minimizes the risk of local pressure peaks in the cable during a phase change of the nitrogen. Attention must hereby be paid to ensuing that a smaller mass flow is set than in the rated operation in order to minimise the heat introduced into the system by the pumping process.

The nitrogen begins to evaporate at the point where it reaches the temperature of the bubble-point curve. This will increase the pressure in the phase separator (see Figure 3.24) (red) and can be used to displace the liquid nitrogen in the reservoir (blue). This allows the nitrogen in the cable (green) to be renewed with the saturated nitrogen ($T \approx 77$ K) from the reservoir. With a capacity of more than 5000 kg LN2, this doubles the time t_1 from equation 3.43. This of course requires a corresponding discharge of the already heated nitrogen in the separator and the use of a valve for pressure regulation in the reservoir.

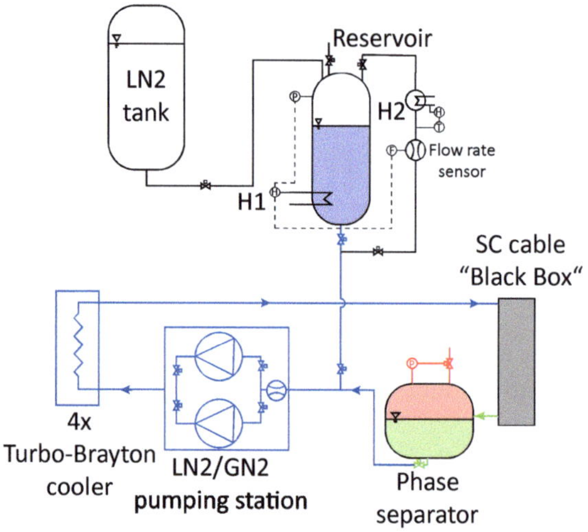

Figure 3.24: Schematic representation of the renewal of the liquid nitrogen in the event of a failure of the chiller. Warm gaseous nitrogen is shown in red, warm liquid nitrogen in green and cold liquid nitrogen in blue .

3.5.3.2 Complete evaporation of the liquid nitrogen

After the time t_1, the nitrogen starts to evaporate. The pressure relief devices must respond and allow the nitrogen to evaporate at atmospheric pressure. The evaporation mass flow is required for the calculation of the evaporation time; this is calculated using the evaporation enthalpy Δh_V at 1 bar (200 J/kg), as

$$\dot{m}_{LN2} = \frac{q_{rad} \cdot l}{\Delta h_V} \approx 0{,}04\ g/s. \tag{3.44}$$

This results in the time until the liquid nitrogen has completely evaporated:

$$t_2 = \frac{m_{LN2}}{\dot{m}_{LN2}} \approx 38\ \text{Stunden} \tag{3.45}$$

3.5.4 Planned heating for maintenance work

The cooling procedure in sub-step 1 (see section 3.5.1) can be used in the opposite direction to heat up the HTS cable. The cable is gradually heated to the operating temperature with gaseous nitrogen. This avoids the formation of excessive axial and radial temperature gradients, also in the planned heating process. The time for heating t_3 can be estimated based on the value that would be needed for cooling ($t_3 \approx 150$ hours). Before the heating with gaseous nitrogen can take place, the liquid nitrogen must be removed.. This is done by feeding the liquid nitrogen into the LN2 tank behind the pump or alternatively behind the Turbo Brayton coolers, depending on the temperature levels at which LN2 is required. The remaining amount of liquid in the system is later flushed out with the gaseous nitrogen.

Heating can also be carried out without gaseous nitrogen by exploiting heat radiation and heat conduction from the ends. In [YMW08[8]] three weeks are required for the "natural" heating process of a 350 meter long HTS cable. This process would lead to longer warm-up times for the HTS cable examined in this report and should be avoided. Heating-up with gaseous nitrogen is preferable in this case.

[8] https://doi.org/10.1063/1.2908453

3.6 Own requirement for the cooling

As indicated in Table 3.24, a superconducting cable system on the sample route requires an output of about 1 MW for the highest operating current. This service shall be provided at the location where the chiller is located.

Cable systems in the transmission network are normally not directly connected in a substation which already provides a high supply capacity to cover its own needs. Rather, the power must be provided separately in the cable transition system.

Power voltage transformers (Power VTs), which can be directly connected to the 380 kV line, are generally used in a cable transition system. They are available on the market as single-phase units for a supply of up to 100 kW and thus cover the requirements of a conventional cable transition system. In order to achieve the higher power required, the devices must be further developed to reduce the number of Power VTs required and thus the space required. Alternatively, own requirements can be covered by the construction of a dedicated medium-voltage line, for example from the regional supply network.

In the event of a failure of the own supply or the absence of the feeding grid, a corresponding redundancy must be provided. In the AC supply sector, this can consist of an emergency power generator, for example. A battery system must be installed for the uninterrupted supply of important DC consumers. If the system is to be designed for cold-startup capability, sufficient fuel must be provided.

4 Economic efficiency

4.1 Introduction

4.1.1 Calculation of the annual energy loss

Cables are not constantly loaded with the same transmission capacity. Rather, the load fluctuates depending on the respective grid situation. The term annual load factor m_a is introduced in order to indicate the ratio between the energy used and the peak load multiplied by time. Equation 4.1 shows the annual load factor m_a.

$$m_a = \frac{\frac{1}{t_a}\int_0^{t_a} S(t) \cdot dt}{S_N \cdot t_a} \tag{4.1}$$

The result for one year is $t_a = 8760$ Stunden. Figure 4.1 shows the curve of the grid load.

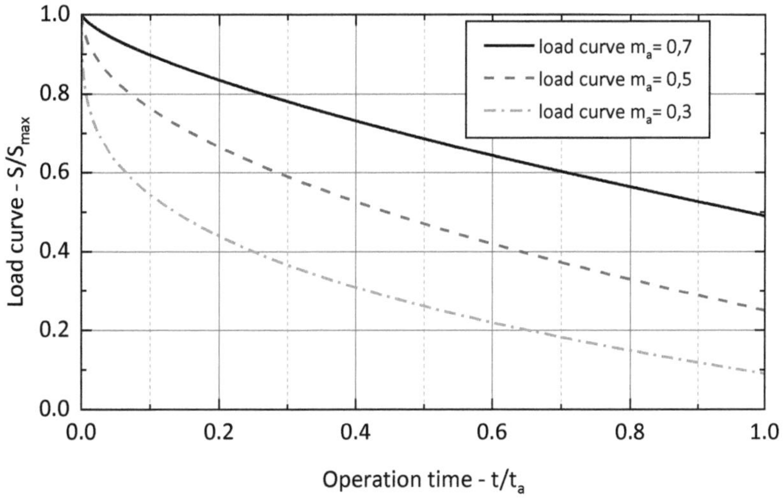

Figure 4.1: Theoretical load curve according to Junge for calculating the energy loss factor with a load factor of $m_a = 0.3$, $m_a = 0.5$ und $m_a = 0.7$.

When the load factor is used, the individual losses are calculated with the maximum load, for example the maximum amperage. The load curve y is specified according to equation 4.2 [Zeb59].

$$y(x) = 1 - (1 - m_a^2) \cdot x^{m_a} \tag{4.2}$$

The load curve y is converted into a so-called energy loss factor ϑ using equation 4.3. The ratio between time t and annual operating time t_a is abbreviated with the variable $x = t/t_a$. This makes it possible to calculate the annual energy loss for current-dependent losses.

$$\vartheta = \int_{x=0}^{x=1} y^2(x) \cdot dx \tag{4.3}$$

If equation 4.2 is used in equation 4.3 and integrated beyond the given limits, the result is simplified equation 4.4.

$$\vartheta = \frac{m_a^2 \cdot (2 + m_a^2)}{1 + 2 \cdot m_a^2} \tag{4.4}$$

For the energy loss factor ϑ_{SL} of the superconducting cable, a dependence on the third power of the current is assumed in the first approximation. This results in the integral presented in equation 4.5 [internal study, "Ampacity", 2010].

$$\vartheta_{SL} = \int_{x=0}^{x=1} y^3(x) \cdot dx \tag{4.5}$$

Integration beyond the given limits results in equation 4.6.

$$\vartheta_{SL} = \frac{m_a^3\left((2 \cdot m_a^2 + m_a + 3) \cdot m_a^2 + 6\right)}{m_a \cdot (6 \cdot m_a + 5) + 1} \tag{4.6}$$

The loss components that cause direct electrical losses are included in the annual energy losses calculation. Thermal losses are compensated and calculated via the cryocooler or the nitrogen cooling costs.

Table 4.1: Overview of the calculation equation for various annual energy losses of operating resources

Variations	Calculation
Conventional underground cable	$W_{\mathrm{V,g}} = \left(P_{\mathrm{V,L}} \cdot \vartheta + P_{\mathrm{V,0,D}}\right) \cdot t_a$
Superconducting cable	$W_{\mathrm{V,g}} = \left(\underbrace{P_{\mathrm{V,SL}} \cdot \vartheta_{\mathrm{SL}}}_{\text{SL-Kabel}} + \underbrace{P_{\mathrm{V,th,SZ}} + P_{\mathrm{V,el,SZ}}}_{\text{Stromzuführung}} \cdot \vartheta + \underbrace{P_{\mathrm{V,0,LK}}}_{\text{Leitungskryostat}} + P_{\mathrm{V,0,D}} \right) \cdot t_a \cdot \eta^{-1}$

$W_{\mathrm{V,g}}$	Total energy loss	(Wh)
$P_{\mathrm{V,SL}}$	Alternating current losses	(W)
$P_{\mathrm{V,0,LK}}$	Cryostat transmission losses	(W)
$P_{\mathrm{V,SZ}}$	Power supply losses	(W)
$P_{\mathrm{V,0,EK}}$	Cryostat termination losses	(W)
ϑ	Energy loss factor for conventional components	
ϑ_{SL}	Energy loss factor for superconducting components	
t_{a}	Annual operating times	(h)
η	Efficiency of the cryocooler	

It should be noted that for superconductor cables with cold dielectric, the dielectric losses must be compensated by the cooling. Table 4.2 indicates the annual energy losses of the two intermediate cabling variants.

Table 4.2: Overview of the annual energy losses for two circuits with different load factors

Variations/ Annual energy loss	Symbol	Annual energy loss $m_a = 0{,}3$ MWh	Annual energy loss $m_a = 0{,}5$ MWh	Annual energy loss $m_a = 0{,}7$ MWh
Conventional underground cable				
Active-power losses	$W_{V,L}$	1894	4455	7320
Dielectric losses	$W_{V,0,D}$	1189	1189	1189
Annual energy loss – total	$\boldsymbol{W_{V,g}}$	3082	5643	8509
Super conducting cable				
Alternating current losses	$W_{V,SL}$	103	321	701
Thermal power supply	$W_{V,th,SZ}$	100	100	100
Electric power supply	$W_{V,el,SZ}$	29	68	112
Line cryostat	$W_{V,0,LK}$	2670	2670	2670
Termination cryostats	$W_{V,0,EK}$	33	33	33
Dielectric losses	$W_{V,0,D}$	828	828	828
Annual energy loss – total	$\boldsymbol{W_{V,g}}$	3763	4020	4444

In Figure 4.2 the annual energy losses of the two intermediate cabling variants are shown as a function of the load factor m_a.

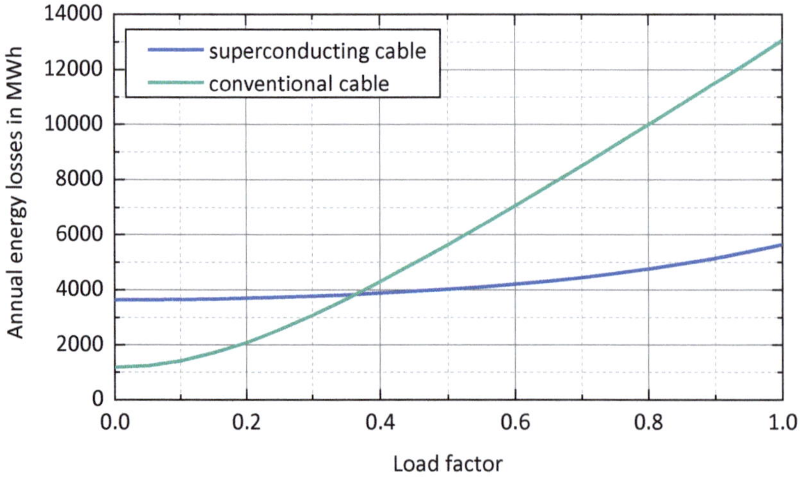

Figure 4.2: Overview of the annual energy losses of the various intermediate cabling variants as a function of the load factor for a route length of 3.2 km.

From an annual load factor of approx. 0.35, the superconducting cable has a lower energy loss than the conventional underground cable.

4.1.2 Method of calculation

The basic idea of the present value method is to convert all costs incurred during a project to the present value. All operating costs incurred over the useful life are discounted. The discounting S_{Ab} indicates how much capital should be profitably invested or invested today in order to be able to pay the operating costs in the coming years. In business practice, capital investment does not take the form of securities, but of investments in one's own company and the resulting profits. The **WACC interest rate** q specifies a required minimum interest rate for this capital-increasing interest income.

Equation 4.7 is used to calculate the discount rate S_{Ab} from the sum of the annual operating costs multiplied by the imputed interest rates q^{-i}.

$$S_{\mathrm{Ab}} = \sum_{i=1}^{N} K_{0\mathrm{i}} = \sum_{i=1}^{N} K_{\mathrm{i}} \cdot q^{-i} \tag{4.7}$$

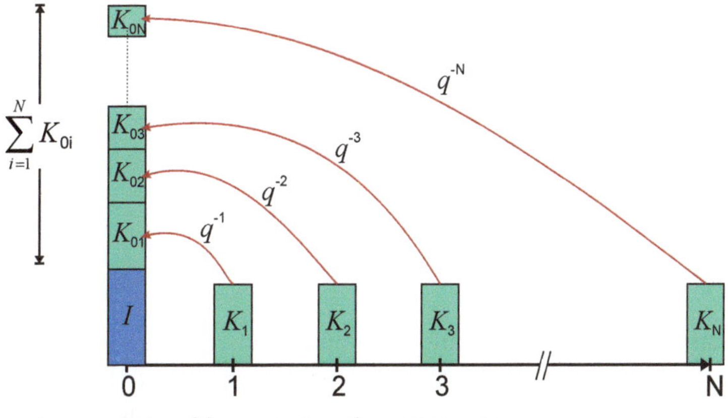

Figure 4.3: Representation of the summation of operating costs by discounting over the service life of N years.

It is assumed that the annual operating costs $K_{\mathrm{i}} = K$ are the same over the entire period of use N - in this study for the assumed operating period of 40 years - and that the interest rate $q_{\mathrm{i}} = q$ remains constant, the present value can be determined using the simplified equation 4.8. The investment costs I are added to the discounting.

$$B = I + K \cdot \underbrace{\frac{q^N - 1}{q^N(q-1)}}_{r_B} \tag{4.8}$$

The last term of equation 4.8 is called the present value of annuity factor $r_B \approx 19{,}5$ and is used to simplify further calculation.

4.1.3 Basic assumptions

The cost calculations for the economic feasibility study are based on the assumptions made in Table 4.3. It should be noted that a detailed real economic study, with interest rate fluctuations and tax accounting, is not the aim of this study. Therefore, only a single calculation with the assumed numbers is made as example. It is also assumed that all circuits or their power systems are symmetrically loaded when possible.

Table 4.3: Parameters for Cost Accounting

Label	Values
WACC interest rate	4.1 %
Observation period	40 years
Annuity factor	19.5
specific costs of losses (internal power supply[a)])	30 €/MWh
specific costs of losses (the external supply[a]) [STA18]	172 €/ MWh

Internal and external supply from the perspective of the transmission network operator, described in the introduction to section 3.5.

4.2 Costs

The costs arising from the installation and operation of line systems are essentially made up of the investments made at certain points in time and the annual operating costs. (Figure 4.4). Other costs, e.g. repair costs, occur stochastically, e.g. in the event of disruptive events with damages. As they are difficult to estimate and are subject to great uncertainty, also due to a lack of statistical information, they are not taken into account. Loss costs account for the largest share of operating costs.

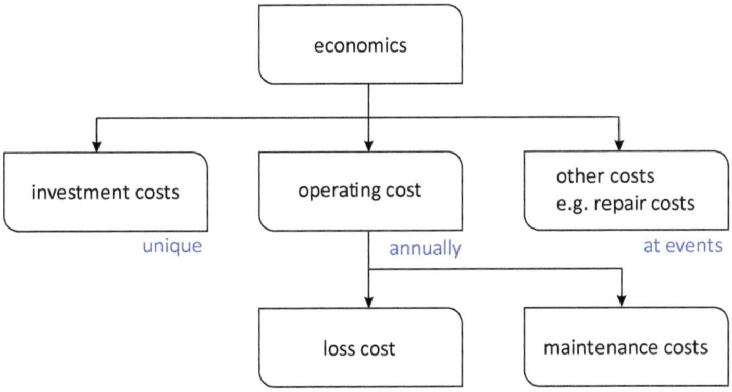

Figure 4.4: Cost shares for installation and operation of line systems.

4.2.1 Investment costs

The investment costs listed below are based partly on information provided by the manufacturer and partly on practical cost estimates. These costs can be taken as rough guidelines. Especially the price of the SC tape fluctuates strongly depending on the manufacturer.

Table 4.4: Investment costs for a super conducting cable of both systems on a route length of 3.2 km

Resources	Low costs T€	High costs T€	Specification
Superconducting cable	20,690	34,520	
Material costs HTS	9,320	18,640	YBCO -SC tape 4 mm, Ic=150 A, 25-50 €/kA/m, 2 system 2485, 2 km
Insulation material Tyvek™	815	1,548	DuPont Tyvek™, 100 m foil 75 g/m², cost 20-38 €/kg, density 340 kg/m³, volume per cable 20.1 m³, mass per cable 6800 kg, cost per cable 258 k€
Line cryostat with weld coupling	8,674	9,830	CryoTherm, rigid cryostat tube DN 100, 450-510 €/m, 12 m single length, 6 x 3200 m
Production costs	1,881	4,503	Assumption: Labour costs correspond to approx. 10-15% of material costs
Chiller	4,400	5,750	
Cryocooler	4,000	5,000	Air Liquide, 4 units TBF-175, cooling line 16.5 kW @ 77 K, € 1.0–1.3 million per unit
Chiller equipment	400	750	Assumption: Accessories costs correspond to approx. 10-15 % of the cryocooler costs.

Sets	3,100	4,650	
Termination	1,200	1,800	Nexans, single-phase terminations, 100-150 T€, 6 pieces (internal communication)
Sleeves	1,900	2,850	Nexans, sleeves, 50-75 T€/piece, each approx. 500 m one sleeve (internal communication), 38 pieces
Civil engineering	3,040	4,480	Assumption: civil engineering costs halved, with approx. halved excavation costs
Cable trench	1,120	1,600	350-500 k€/km [Osw07]
Wiring duct	1,600	2,400	500-750 k€/km [Osw07]
Laying of cables	320	480	100-150 k€/km [Osw07]
Total system costs	31,230	49,400	

The investment costs for two superconducting insulated cables amount to € 31 to 49 million. At € 21 to 34 million, cable costs account for around three quarters of total costs.

Table 4.5: Investment costs for a conventional underground cable with four systems on a route with length of 3.2 km

Resources	Low costs T€	High costs T€	Specification
Conventional underground cable	28,800	43,200	Four parallel 380 kV three-phase current cable systems 3 x 1 x 2500 mm² 2XS(FL)2Y in two trenches incl. laying, routing, fittings, tests etc. 9,000–13,500 k€/km

The comparison between conventional cables and superconducting cables show that with low cost assumptions, both insulated cables have similar investment costs.

4.2.2 Cost of losses

The annual energy loss$W_{V,g}$ multiplied by the loss energy price k_{VW} results in the loss energy costs K_V according to equation 4.9. The decisive factor here is how the connected load of the refrigeration system of the superconducting cable is provided: If the loss energy is drawn from the distribution network, the operator of the extra-high voltage cable system behaves like an industrial customer of the distribution network operator. In this case, a loss energy price of 172 €/MWh is assumed (external supply from the point of view of the transmission system operator). If the transmission system operator is prepared to supply the chiller or the energy loss, a loss energy price of 30 €/MWh is assumed according to Table 4.3 (internal power supply from the transmission network operator's point of view).

$$K_V = W_{V,g} \cdot k_{VW} \tag{4.9}$$

In Table 4.6 the annual lost costs are listed.

Table 4.6: An overview of the annual loss costs

Variations	Internal power supply through transmission system operator			External supply through Distribution network provider		
	Cost of losses $m_a = 0,3$ €k/year	Cost of losses $m_a = 0,5$ €k/year	Cost of losses $m_a = 0,7$ €k/year	Cost of losses $m_a = 0,3$ €k/year	Cost of losses $m_a = 0,5$ €k/year	Cost of losses $m_a = 0,7$ €k/year
Conventional Underground cable	92.5	169.3	255.3	-	-	-
Super conducting cable	112.9	120.6	133.3	647.2	691.4	764.4

4.2.3 Maintenance costs

The maintenance and repair costs for superconducting cables and conventional underground cables are estimated to be similarly high. For this reason, maintenance costs are not considered.

4.2.4 Present value of total costs

Figure 4.5 shows a representation of the total present values as a function of the load factor, in the case of internal supply of the power loss by the transmission system operator.

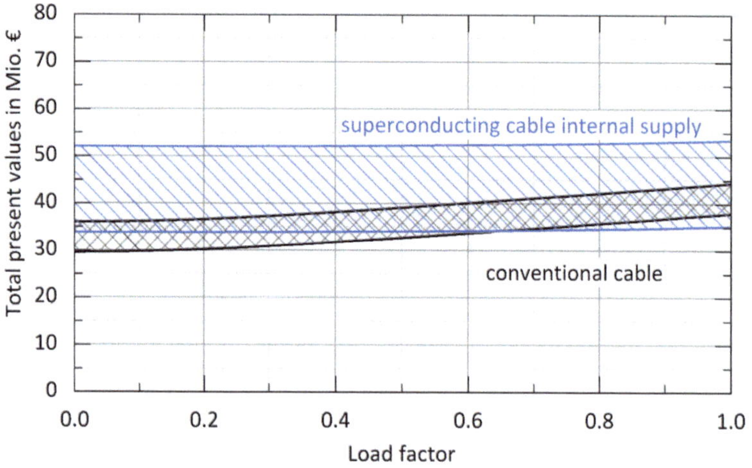

Figure 4.5: Overview of the total present values as a function of the load factor, consisting of investment costs and operating costs over a service life of 40 years.

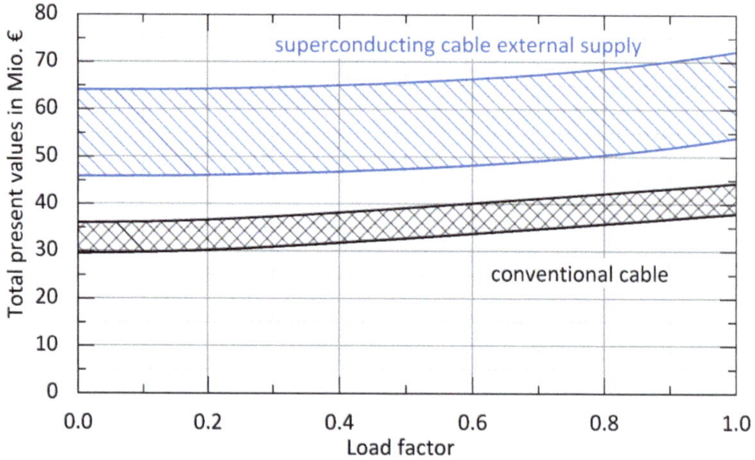

Figure 4.6: Overview of the total present values as a function of the load factor, consisting of investment costs and operating costs over a service life of 40 years..

Already in open-line operation of the superconducting cable there is a cost overlap between the transmission technologies to be compared. This means that depending on the actual price development of the cable system components, the system costs for a superconducting cable may be lower than the system costs for a conventional underground cable. From a load factor of 0.6, the cost ranges of superconducting and the conventional cable system overlap completely. The complete overlapping of the cost ranges implies that the superconducting cable must in any case be considered as an economic alternative. Figure 4.6 shows an overview of the total present values as a function of the load factor for the case of external supply for power loss.

If the losses of the superconducting cable system are compensated by a regional distribution grid operator, this is considered to be external supply from the point of view of the transmission grid operator. Depending on the load factor, there is no cost overlap for this type of supply. The system costs for the superconducting cable are always higher than the system costs of the conventional underground cable. In Figure 4.7 the total present values of the tested cable systems are shown with a load factor of 0.35. The total present values are broken down into operating and investment costs.

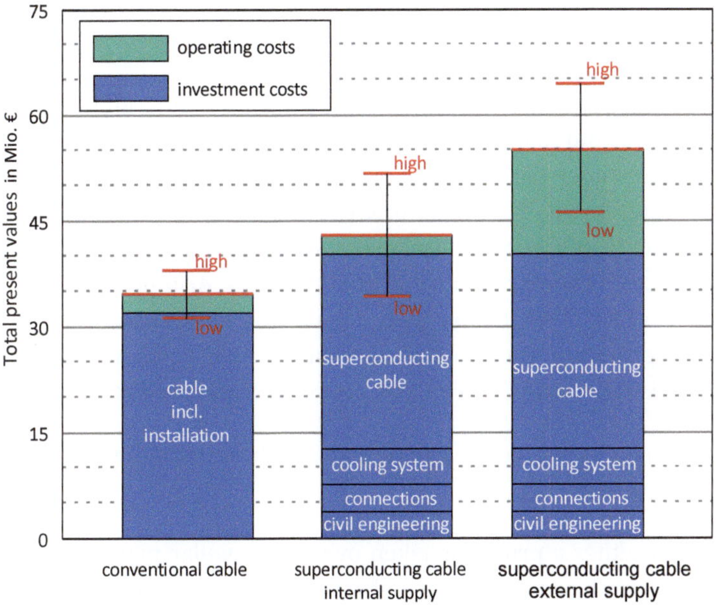

Figure 4.7: Overview of the total present values as a function of the load factor 0.35 divided into investment costs and operating costs over a service life of 40 years.

In addition to information on operating and investment costs, each of the three present value bars contains a red tolerance range. This tolerance range is already described in detail in Table 4.4 and is used to estimate the cost development fluctuation in superconducting cable technology.

With conventional cable and with superconducting cable with internal power supply (from the viewpoint of the transmission network operator), the low impact of the operating costs is evident. In the case of external supply (from the point of view of the transmission system operator), however, the operating costs correspond to approx. 25 % of the total present values. This supply type is therefore not recommended.

4.3 Regulatory framework

This section explains whether and to what extent the operating costs of a superconducting cable can be recognised by the network operator under the currently applicable regulatory conditions.

4.3.1 Investment costs (CAPEX)

In principle, expansion measures involving superconductors can be approved as investment measures pursuant to Section23 Incentive Regulation Ordinance (ARegV), as they fall either under the catalogue of Section 23 (1) sentence 2 ARegV or under the catch-all requirement of "demand-oriented expansion" pursuant to Section 11 Energy Industry Act (EnWG). Recognisable costs include both cable transition systems and additional costs compared to conventional cables. Also possible could be an interpretation as quick construction costs, which may be necessary in order to complete a specific project on time. In this marginal case, the costs would be regarded as incidental acquisition costs, otherwise as acquisition and production costs (APC).

If a project is exclusively used for scientific research, recognition pursuant to Section 23 ARegV is not possible.

Regardless of whether a project is taken over into regular network operation or is exclusively a research project - there is also the possibility of recognition via the base year system. However, only the cost level of the base year (less proportionate inefficiency) would be eligible for recognition, which is why the time of activation may have to be taken into account during implementation. In addition, there is a fundamental risk that the German Federal Network Agency will make a reduction or demand a special justification of the costs.

Aside from this, a recognition via section 25a ARegV for research projects is also permissible in principle for capital costs. However, only 50 % of the costs can be allocated to the network charges. An essential prerequisite for this is funding in a project of a federal or state ministry.

4.3.2 Operating costs (OPEX)

Annual operating costs can be allocated using the various instruments listed below.

4.3.2.1 Recognition of the investment measure during the approval phase

In the event of recognition pursuant to Section 23 ARegV, 0.8% of the APC activated at the time can be claimed as operating costs. The approval period is limited to the end of the regulatory period in which the project is completed. Depending on the project layout, this can already affect a regular operating phase for a superconductor section. Other operating costs and costs after the approval period can only be allocated through other instruments.

4.3.2.2 Recognition via voluntary self-commitment grid losses

An recognition of the operating costs via the voluntary self-commitment (VSC) grid losses and thus the categorisation as permanently uninfluenceable costs in accordance with Section 11 (2) S. 2-4 ARegV is neither currently possible nor planned for the coming regulatory period. The voluntary self-commitment concerns only the procurement of network losses and not opportunity costs in the form of operating costs of a superconducting cable section.

4.3.2.3 Recognition via the base year cost audit

The operating costs can be recorded via the cost check. In this calculation, only the cost level of the base year (less proportionate inefficiency) can be allocated. For an exclusively research project this would be the priority variant of recognition.

4.3.2.4 Recognition within the framework of a research project

Recognition of the operating costs would also be possible in a subsidiary manner through Section 25a Incentive Regulation Ordinance (ARegV). A maximum of 50 % of the costs can be recognised here. The funding of a federal or state ministry is a necessary prerequisite for this recognition.

4.3.2.5 Regulatory implications

In principle, there is a high risk that all categories of additional costs will have a negative effect on the efficiency rating. In European benchmarking, which is not binding in Germany for the third regulatory period - 2018 to 2022 - this effect would be largely offset by the type of output included (normalized grid). In a reference network analysis, this depends on the concrete model design.

In the variant currently under discussion, additional costs due to superconductors would probably not lead to disadvantages. This design has not yet been fully determined. A disadvantage of such effects is still possible for the third period, which then also applies in the coming periods if the model remains constant.

5 Failure behaviour of superconducting 380 kV cables

5.1 Fault scenarios

Any electrical equipment can be affected by faults and thus be irreparably damaged.

With conventional cable systems, a distinction must be made between faults in the insulation that immediately lead to a short circuit and other defects. Defects in the ancillary systems usually do not mean that the system must be switched off immediately, but can in many cases be repaired during operation of the cable system. An insulation fault in the cable, a sleeve or a termination leads to an immediate failure of the cable system. The error must then be corrected as described in the section 5.3.

Table 5.1 describes the main faults and their effects as well as suitable remedial measures for superconducting cables.

Table 5.1: Significant potential faults in superconducting cables and their effects

Error	Impact and measures	Remedy
Breakdown in the dielectric	Disconnection and replacement the affected cable section	Careful design, manufacture and testing as well as TE monitoring
Current overload	Quench of the superconductor and, if the design parameters are exceeded, disconnection of the cable route.	Design and installation of secondary protection technology
Failure of the power supply	Failure of cooling and sensors, after some time the cable route is switched off.	Battery storage for sensors and UPS for chiller or orderly shutdown of the system
Failure of cooling	After some time the cable section is shut down.	Redundancy through modular cooling system
Break of the vacuum in the cryostat	Disconnection and replacement the affected cable section	Regular control and clear marking of the cable laying
Faulty sensors	Continued operation with fast checking, troubleshooting or shutdown, depending on the criticality of the sensor.	Redundancy due to different sensor types and measuring locations

In most cases, the cable section can be switched off in an orderly manner. A time-consuming replacement of the affected cable section only becomes necessary in the case of permanent damage to the cable in the superconductor, the dielectric and the cryostat. However, this case has not yet occurred in the operation of superconducting cables.

Automatic reconnection after a short interruption does not affect a superconducting cable. Since the thermal processes take place with a very high time constant of several minutes as a rule, a superconducting cable is not affected. With the "AmpaCity" cable in Essen, this case was tested, and with future short-term interruptions and with automatic reclosing the cable no longer exists the grid.

5.2 Fault frequency

For conventional cable systems, the error frequency is summarised in the CIG09 study. The error rate for three-phase high-voltage cables is given as 0.133 errors per year and 100 system kilometres. 0.048 defects per year and 100 pieces were determined for sleeves, 0.050 defects per year and 100 pieces for terminations. It should be noted that the database in the study was very small due to the small number of extra-high voltage systems in operation and that the error frequency in cable systems can deviate significantly from these values.

To evaluate a fault frequency and thus the reliability and availability of superconducting insulated cable, it takes many years of operating experience from which data such as MTBF (Mean Time Between Failures) or a percentage availability can then be derived. To date, most development projects have aimed to demonstrate the technical feasibility and the use in the power grid. Thus, these statistical data for superconducting cables are not yet collected. Nevertheless, some quantitative assessments can be made.

None of the network demonstrations of superconducting cables have so far reported a degradation of the cable. Ageing and faults in the cable can always occur in the electrical insulation, the superconductor and the cryostat.

Since the electrical insulation is a liquid-solid mixture of liquid nitrogen and polypropylene laminated paper (PPLP), there is basically a good strength even against partial discharge voltages; the dimensioning is similar to the conventional procedure. High-temperature superconductors are not known to age as long as they are operated below their critical current and the maximum mechanical stresses are maintained. Thus, a degradation would only be conceivable through inadmissible mechanical stress or current load, which must be avoided by the operator and appropriate protective measures, such as shutdown in the event of faults. A loss of the vacuum in the cryostat would also be conceivable. A review is therefore recommended after ten years of operation.

This means that it can be assumed that the reliability of a superconducting cable system is determined mainly by the reliability and availability of the cooling system. The chiller systems are already designed with redundancy in the demonstration projects. In the "AmpaCity" cable in Essen, for example, the main redundancy is in the pumps for the circulation of liquid nitrogen. These were regularly maintained from the beginning. In this project, if one pump fails another one takes over the function and the affected part can be replaced and maintained without shutting down the chiller. The cable has been in continuous operation since its official commissioning in April 2014. The cable in Essen is thus fully integrated into the operation of the network. For individual chillers such as Air Liquide's Turbo Brayton system, MTBF values of 10,5000 hours and five-year maintenance intervals are available. The impact of this on overall availability has not yet been investigated.

5.3 Repair scenario

For a conventional plastic-insulated cable, a total failure time of three to four weeks can be expected in the event of a cable or socket fault. During this period, the following phases are run through:

- Fault locating
- Construction preparation and civil engineering for uncovering cables
- Cutting the cable
- Removing the defective cable piece
- Installation of a replacement cable
- Production of sleeve connections
- Dismantling
- Inspection

For a superconducting cable, the individual phases are exactly the same, but the cable must be heated before being cut and re-cooled after the sleeve connection has been made. This extends the downtime by an estimated one to two weeks depending on the cable length, whereby the warming up can be accelerated by introducing gaseous nitrogen. Since the superconducting cable has been divided into individual sections between the sleeves, it is recommended to replace one section. In principle, however, partial sections can also be replaced by inserting further sleeves.

5.4 Spare parts concept

With conventional plastic insulated cables, a fault in a sleeve or on the cable can usually be repaired with a short replacement piece of a few meters. If there is a cable fault in the area of a crossing, a replacement cable of at least the length of the hole must be used; this can be several hundred metres. Corresponding cable lengths must be kept available for short-term repair work.

In addition to the cable spare part, other spare parts such as terminations, sleeves, other system parts and installation material must be kept in stock.

Since the cooling system for superconducting cables is redundant, it is necessary to check delivery times to see whether spare parts stocking or procurement is the more economical option. Otherwise, cable lengths, terminations, sleeves and other plant components and installation material must be kept available as is the case with conventional cables, .

6 Summary evaluation

This study compared a 3.2 km long superconducting 380 kV cable with a design current of 3.6 kA with conventional underground cables from a technical and economic point of view. The study did not make a comparison with an overhead line, as there are already several extensive studies on the comparison of overhead and underground cables.

Superconducting AC cables have so far been developed for a voltage of 275 kV and are also technically feasible for voltages of 380 kV. The necessary technologies are scalable. Partial underground cables with lengths of less than 5 km are often planned on new construction pipelines. This section length can be achieved with superconducting cables without separate intercooling.

Compared to conventional systems with plastic-insulated cables, superconducting systems have the advantage of a narrower route. The electrotechnical parameters of a superconducting cable system are more similar to those of an overhead line than those of a conventional cable system, which offers advantages for network technology and operational management. However, the maintenance overhead is higher due to the active cooling. It should be noted that superconducting cable systems have limited cold-startup capability. This characteristic must be taken into account in the network planning evaluation of a line.

The comparison of the losses of a superconducting cable system with those of a conventional cable system shows a high dependency on the load: While superconducting systems perform better when the lines are highly utilised, conventional systems have lower losses when lines are less utilised. The expected utilisation rate is therefore of great importance for the decision on the type of cable system.

The study also showed that a superconducting cable system can already be economical today. Economic efficiency is mainly influenced by the material costs for the SC tape, the capacity utilisation of the installation and the costs for providing the electrical energy required for the operation of the cryopumps.

As the next step on the way to the realization of a superconducting 380 kV cable system, a manufacturer would have to review the proposed design and possibly further develop it for production. The cable with the necessary connection sleeves and terminations would then have to be manufactured and installed and then intensively tested together with a cooling system. If necessary, the test may result in adjustment requirements for the components or the entire system. The IEC standard "Superconducting AC power cables and their accessories for rated voltages from 6 kV to 500 kV - Test methods and requirements", which is in preparation at the end of this study, should be observed during development and testing.

List of references

[AAL18] Advancedtech Air Liquide: "Turbo-Brayton cryogenic systems - The cryogenic technological break-through for refrigeration and liquefaction from 25 K to 200 K", https://advancedtech.airliquide.com/turbo-brayton-cryogenic-systems Page view 30.03.2018

[Bay16] C. M. Bayer: "Characterization of High Temperature Superconductor cables for Magnet Toroidal Field Coils of the DEMO Fusion Power Plant", 2016, KIT Scientific Publishing, Volume 18, Page 45, http://dx.doi.org/10.5445/KSP/1000062054

[CIG09] CIGRE Technical Brochure No. 379 – Update of Service Experience of HV Underground and Cable Systems, B1, 2009

[CIG15] CIGRE Technical Brochure No. 644 –Common Characteristics and emerging test techniques for HTS Power Equipment, D1, 2015

[CKK12] Y.S. Choi, D.L. Kim et al.: "Performance Test of Cooling System for 500 m HTS Cable in KEPCO Power Grid", IEEE Transactions on Applied Superconductivity Vol. 22 No.3, (2012), (https://doi.org/10.1109/TASC.2011.2178376)

[CRY12] Cryoflex-Nexans – Manufacturer catalogue – *CYROFLEX transfer lines for liquid gas*, 2012

[CRY15] CryoTherm – Manufacturer catalogue – Cryo cable systems, 2015

[CRY17] H. Chang, K. Ryu, H. Yang: "Cryogenic design of liquid-nitrogen circulation system for long-length HTS cables with altitude variation", In Cryogenic (2017), 83, 50-56, (http://dx.doi.org/10.1016/j.cryogenics.2017.02.004)

[DEN14] Technologieübersicht (technology overview): The German extra-high voltage grid: Technologien und Rahmenbedingungen; Deutsche Energie-Agentur GmbH (dena); Hannes Seidl, July 2014

[Det04] Martin Detert, Cornelia L.: Hydraulik von Rohrsystemen. (Hydraulics of pipe systems) Kurs SS 2004. Faculty of Civil Engineering, Geosciences and Environmental Sciences (BGU) Institute of Hydromechanics

[Deu12] Deutsche Umwelthilfe e.V.: Himmel und Erde – Freileitungen und Erdkabel (Heaven and Earth - Overhead Lines and Underground Cables), Berlin, April 2012

[DIN48] DIN EN 10380. Pipelines - Corrugated metal hoses and hose assemblies

[DIN80] DIN EN 10380. Pipelines - Corrugated metal hoses and hose assemblies, German version EN, 2012

[ESE75] Edison-Sault Electric Co. installed this 138 kV submarine cable in the Straits of Mackinac Summer 1975. Online Access: http://atlantic-cable.com/Cables/Power/index.htm

[FUJ17] Fujikura, https://www.fujikura.co.uk/products/energy-and-environment/2g-ybco-high-tempurature-superconductors/, Webzugriff 2017

[HCB93] Hermann, F.; Cottevielle, C.; Bock, J.: "Cryogenic load calculation of high TC current lead". In Cryogenic (1993), 33 no. 5, S.555-562

[HKS03] J. Howe, B. Kehrli, F. Schmidt: "Very Low Impedance Superconductor Cables", A white paper, 2003

[HNM14] N. Hayakawa, et al., "*A Novel Electrical Insulating Material for 275 kV High-Voltage HTS Cable with Low Dielectric Loss*", Journal of Physics, Conference Series 507 (2014) 32021

[JJH15] W. Ji, A. Jacobi, Y. He, W. Tao: "Summary and evaluation on single-phase heat transfer enhancement techniques of liquid laminar and turbulent pipe flow", International Journal of Heat and Mass Transfer 88 (2015) Page 735–754

[Kau71] K. Kauder: "*Strömungs- und Widerstandsverhalten in gewellten Rohren*" (Flow and resistance behaviour in corrugated tubes) Faculty of Mechanical Engineering at the Technical University of Hanover, Dissertation, 1971

[KSN17] D. Kottonau, E. Shabagin, M. Noe, and S. Grohmann: "Opportunities for High-Voltage AC Superconducting Cables as Part of New Long-Distance Transmission Lines", IEEE TRANSACTIONS ON APPLIED SUPERCONDUCTIVITY, VOL. 27, NO. 4, JUNE 2017, https://doi.org/10.1109/TASC.2017.2652856

[Nex11] Nexans: Under Ground Cables, Paris 2011, Page 14

[Noe17] Mathias Noe, Karlsruher Institut für Technologie, innere Kommunikation, 2017

[Nor70] W.T. Norris: *Calculation of hysteresis losses in hard superconductors carrying ac: isolated conductors and edges of thin sheets*, J: Appl. Phys. 3 489, 1970, (https://doi.org/10.1088/0022-3727/3/4/308)

[Osw07] B.R. Oswald: "*380 kV Salzburg line: Effects of the possible (partial) cabling of the section Tauern-Salzach*", Hannover 2007, https://www.e-control.at/documents/20903/-/-/acf6020d-f3a1-45ed-a860-7d03a0c0e75b

[SC12] Stirling Cryogenics: Stirling Power Cooler LPC-RL 2, 2012

[SCA14] Scapa Cable Solutions: "Cable protection solutions for a joined-up world" Manufacturer Catalogue, 2014 Scapa Group

[Sha14] E. Shabagin: "Untersuchungen zum Druck- und Temperaturverhalten supraleitender Kabel" (Investigations into the pressure and temperature behaviour of superconducting cables), Bachelor thesis, ITEP(KIT), 2014

[Sha17] E. Shabagin, C. Heidt, S. Strauss, S. Grohmann : Modelling of 3D temperature profiles and pressure drop in concentric three-phase HTS power cables, 2017 Cryogenics (http://dx.doi.org/10.1016/j.cryogenics.2016.11.004)

[Sta17] Hanno Stagge, TenneT, innere Kommunikation, 2017

[STA18] BDEW. Industrial electricity prices* (including electricity tax) in Germany from 1998 to 2018 (in euro cents per kilowatt hour). https://de.statista.com/statistik/daten/studie/252029/umfrage/industriestrompreise-inkl-stromsteuer-in-deutschland/ (accessed on 20 February 2018).

[Str11] T. Strobel: "Simulation des strombegrenzenden Verhaltens supraleitender Kabel", (Simulation of the current-limiting behaviour of superconducting cables) Diploma thesis 2011, KIT, Institut für Technische Physik

[Süd16] Südkabel: Cable System Engineering Seminar 2016, Mannheim 2016

[SZA65] STEKLY, Z. J. J.; ZAR, J. L.: Stable Superconducting Coils. In: IEEE Transactions on Nuclear Science 12 (1965), 3, S. 367–372

[VDI13] VDI.e.V: *"VDI-Wärmeatlas"* (VDI Heat Atlas), Springer-Verlag Berlin Heidelberg, 2013

[Wes11] Beate West: "Supraleiterkabel - Eine Machbarkeitsstudie", (Superconductor Cable - A Feasibility Study) Nexans 2011

[YMA06] J. Yuan, J. Maguire, A. Allais, F. Schmidt: "A Cool-down and Fault Study of a Long Length HTS Power Transmission Cable", AIP Conference Proceedings 823, 782 (2006), (https://doi.org/10.1063/1.2202486)

[Zeb59] M. Zebisch: *Netzverluste – Die Verluste in elektrischen Versorgungsnetzen, ihre Ursachen und Ermittlung* (Network losses - The losses in electrical supply networks, their causes and determination), VEB Verlag Technik Berlin, 1959

List of figures

Table of contents

List of calculations

Table 1.1	/calculation/calculation of loss/conventional cable
Table 1.2	/calculation/calculation of loss/conventional cable
Table 1.3	/calculation/calculation of loss/conventional cable
Table 3.2	/Calculation/Dielectrical_Design_A5_2017_07_19.m
Table 3.3	/Calculation/Dielectrical_Design_A6_2017_07_12.m
Figure 3.1	/Calculation/Design_A5_2017_02_29.m
Figure 3.2	/Calculation/Design_A6_2017_02_29.m
Figure 3.3	/Images/Figure 4.3 - Representation of the defined operating range.m
	/Images/ Figure 4.3 - Representation of the defined operating range.opj
Table 3.6	/Calculation/Pressure_Design_A6_2017_07_12_EK_L_max.m
	/Calculation/Pressure_Design_A6_2017_07_12_BK_L_max.m
	/Calculation/Design_A5_2017_07_19.m
Table 3.7	
Table 3.8	/Calculation/Design_A6_2017_07_12.m
Table 3.11	/Calculation/Thermal_Design_A6_2017_07_12_EK.m
	/Calculation/Thermal_Design_A6_2017_07_12_BK.m
Table 3.12	/Calculation/Thermal_Design_A6_2017_07_12_EK.m
	/Calculation/Thermal_Design_A6_2017_07_12_BK.m
Table 3.13	/Calculation/Thermal_Design_A6_2017_07_12_EK.m
	/Calculation/Thermal_Design_A6_2017_07_12_BK.m
Table 3.14	/Calculation/Thermal_Design_A5_2017_07_19_EK.m
	/Calculation/Thermal_Design_A5_2017_07_19_BK.m
Figure 3.8	/Calculation/Thermal_Design_A6_2017_07_12_EK.m
	/Images/Figure 4.8 - Representation of the nitrogen temperature one-sided cooling concept_v3.opj
Figure 3.9	/Calculation/Thermal_Design_A6_2017_07_12_BK.m
	/Images/Figure 4.9 - Representation of the nitrogen temperature of the two-sided cooling concept_v4.opj
Figure 3.10	/Calculation/Thermal_Design_A6_2017_07_12_BK.m
	/Images/Figure 4.10 - Display of the cable temperature.opj

Figure 3.11	/Calculation/Thermal_Design_A6_2017_07_12_EK_mdot.m /Calculation/Pump_Power_A6_2017_07_12.m /Images/Figure 4.11 - Representation of the maximum route length mass flow one-sided cooling DN 40_v6.opj
Figure 3.12	/Calculation/Thermal_Design_A6_2017_07_12_BK_mdot.m /Calculation/Pump_Power_A6_2017_07_12.m /Images/Figure 4.12 - Representation of the maximum route length mass flow two-sided cooling DN 50_v3.opj
Table 3.15	/Calculation/Thermal_Design_A6_2017_07_12_EK.m /Calculation/Thermal_Design_A6_2017_07_12_BK.m
Table 3.16	Calculator
Table 3.17	Calculator
Table 3.18	/Calculation/Inductance_Capacitance_Design_A5_2017_07_19.m /Calculation/Inductance_Capacitance_Design_A6_2017_07_12.m
Figure 3.14	/Calculation/Thermal_Design_A5_2017_07_19_EK_final.m /Images/Figure 4.14 - Representation of the power loss as a function of the cable length.opj
Table 3.20	/Calculation/Thermal_Design_A5_2017_07_19_EK_final.m
Table 3.21	/Calculation/Thermal_Design_A5_2017_07_19_EK_final.m
Table 3.22	/Calculation/Losses_Design_A5_2017_07_19.xlsx
Table 3.23	/Calculation/Losses_Design_A5_2017_07_19.xlsx
Table 3.24	/Calculation/Losses_Design_A5_2017_07_19.xlsx
Figure 3.15	/Calculation/var_ueberstrom_Wellrohr.m
Figure 3.16	/Calculation/var_ueberstrom.m
Table 3.25	/Calculation/var_ueberstrom_Wellrohr.m /Calculation/var_ueberstrom.m
Figure 4.1	/Calculation/ Load curve_Junge.m /Images/Figure 6.1 - Representation of the load curve.opj
Figure 4.2	/Calculation/Losses_Design_A5_2017_07_19.xlsx /calculation/calculation of loss/conventional cable /Images/Figure 6.3 - Overview of the annual energy loss as a function of the load factor.opj
Table A.1	/Calculation/Thermal_Design_A5_2017_07_19_EK.m /Calculation/Thermal_Design_A5_2017_07_19_BK.m
Table A.2	/Calculation/Thermal_Design_A5_2017_07_19_EK.m /Calculation/Thermal_Design_A5_2017_07_19_BK.m
Table A.3	Calculator
Table A.4	Calculator

A Attachment

A.1 Relation between maximum transmission length and temperature and pressure limits

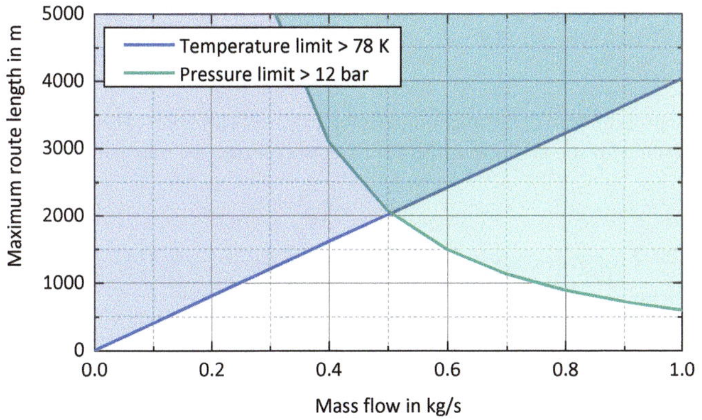

Figure A.1: Representation of the maximum route length as a function of the mass flow with one-sided cooling for the nominal width DN 32 of the corrugated tube concept.

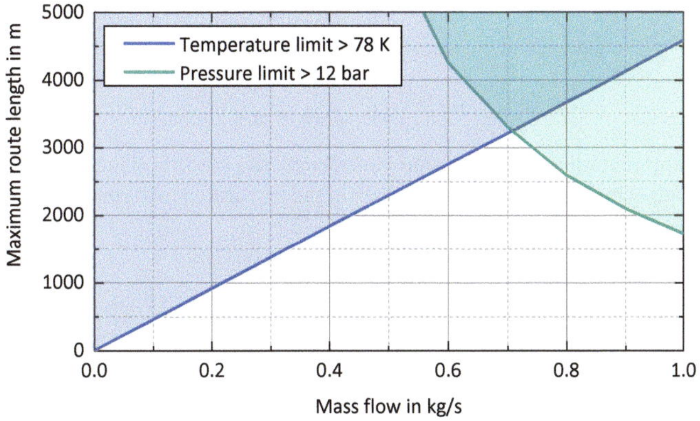

Figure A.2: Representation of the maximum route length as a function of the mass flow with one-sided cooling for the nominal width DN 40 of the corrugated tube concept.

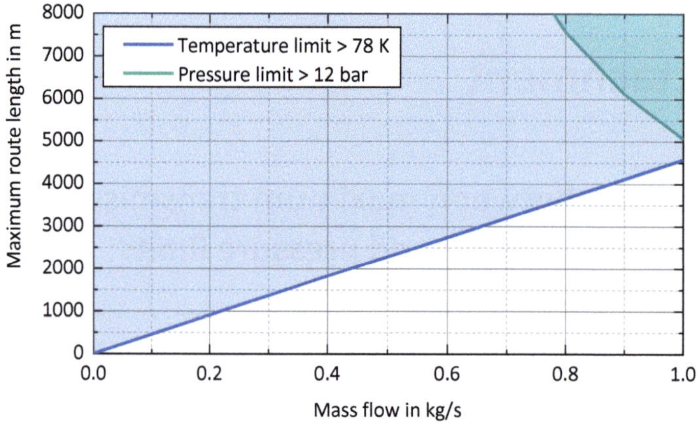

Figure A.3: Representation of the maximum route length as a function of the mass flow with one-sided cooling for the nominal width DN 50 of the corrugated tube concept.

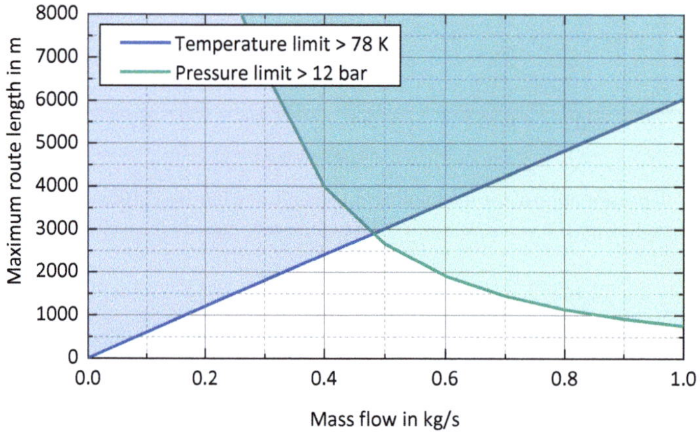

Figure A.4: Representation of the maximum route length as a function of the mass flow with one-sided cooling for the nominal width DN 32 of the corrugated tube concept.

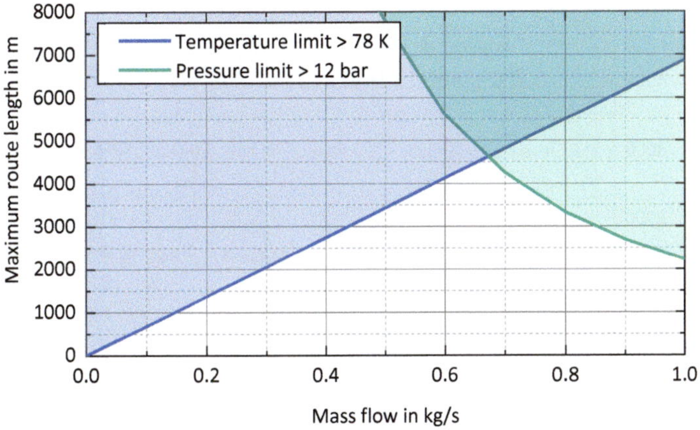

Figure A.5: Representation of the maximum route length as a function of the mass flow with one-sided cooling for the nominal width DN 40 of the corrugated tube concept.

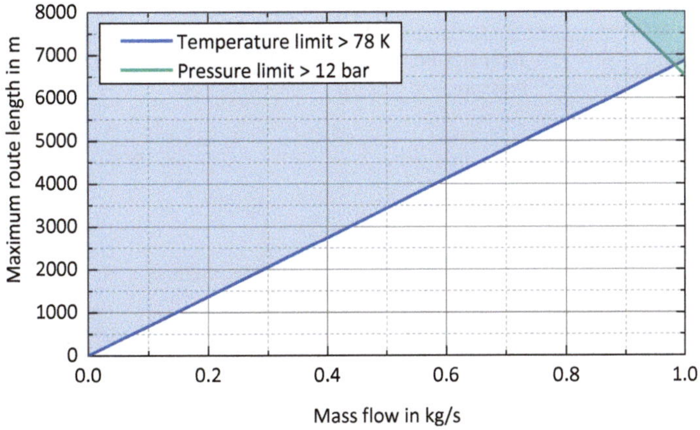

Figure A.6: Representation of the maximum route length as a function of the mass flow with one-sided cooling for the nominal width DN 50 of the corrugated tube concept.

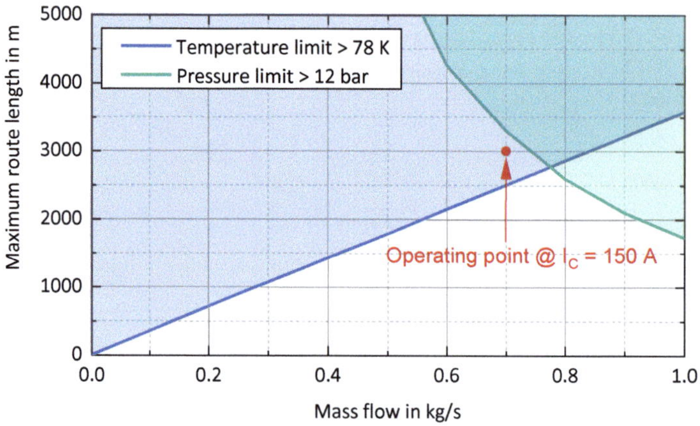

Figure A.7: Representation of the maximum route length in dependency of the mass flow with one-sided cooling for the nominal diameter DN 40 of the inner pipe Ic = 200 A with omitted conductor layer.

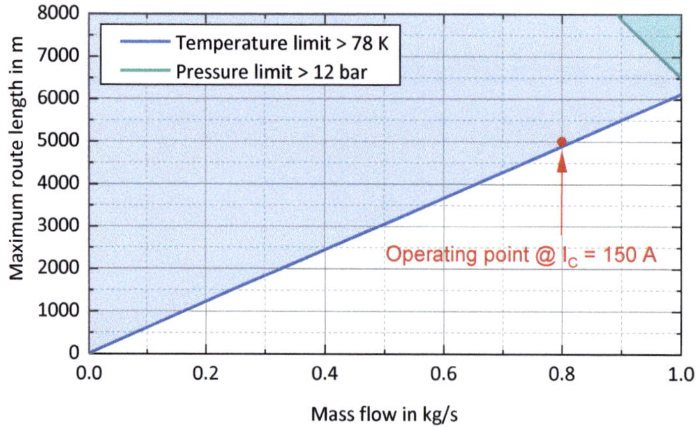

Figure A.8: Representation of the maximum route length in dependency of the mass flow with cooling on both sides for the nominal diameter DN 50 of the inner pipe Ic = 200 A with omitted conductor layer.

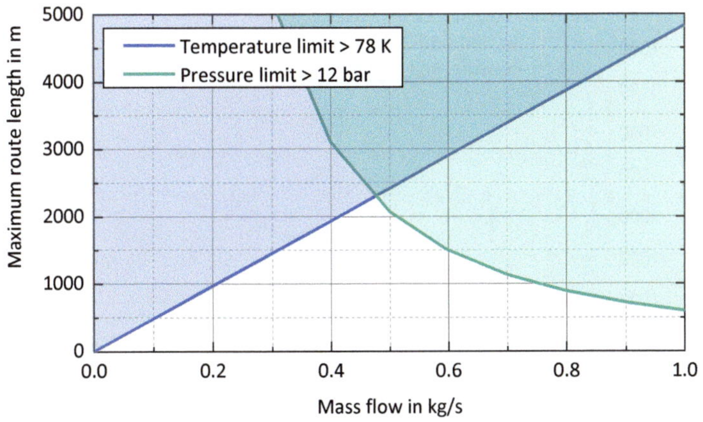

Figure A.9: Representation of the maximum route length as a function of the mass flow with one-sided cooling for the nominal width DN 32 of the hollow core conductor concept.

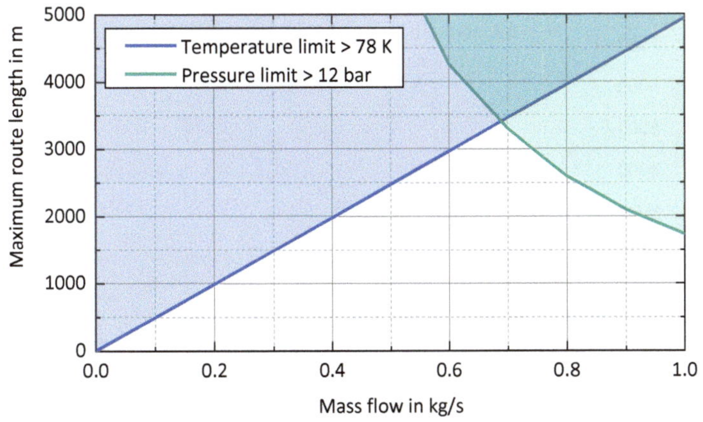

Figure A.10: Representation of the maximum route length as a function of the mass flow with one-sided cooling for the nominal width DN 40 of the hollow core conductor concept.

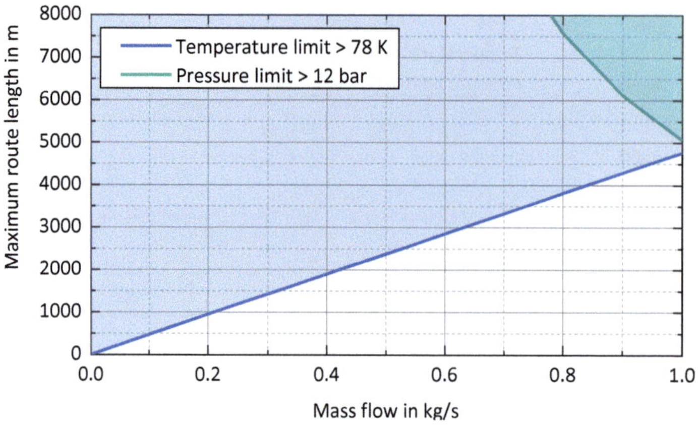

Figure A.11: Representation of the maximum route length as a function of the mass flow with one-sided cooling for the nominal width DN 50 of the hollow core conductor concept.

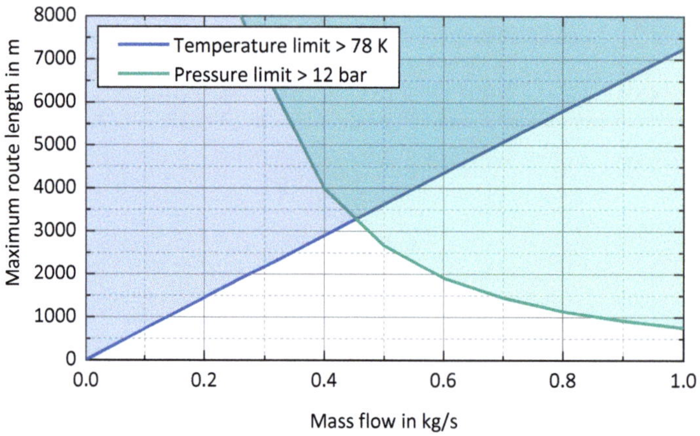

Figure A.12: Representation of the maximum route length as a function of the mass flow with one-sided cooling for the nominal width DN 32 of the hollow core conductor concept.

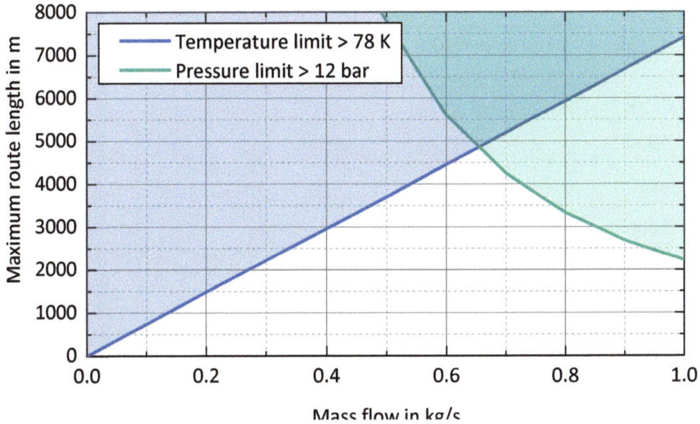

Figure A.13: Representation of the maximum route length as a function of the mass flow with one-sided cooling for the nominal width DN 40 of the hollow core conductor concept.

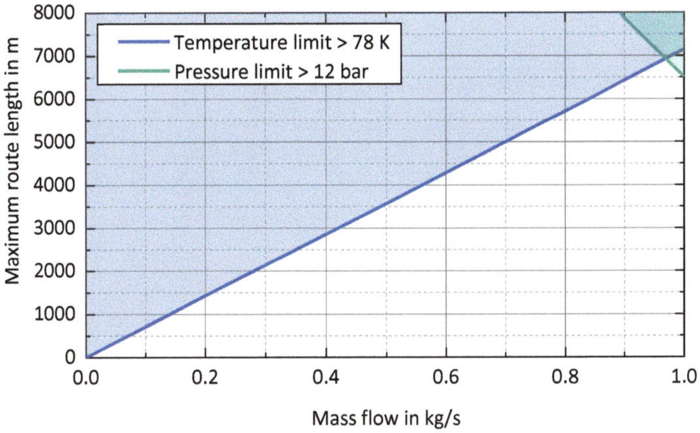

Figure A.14: Representation of the maximum route length as a function of the mass flow with one-sided cooling for the nominal width DN 50 of the hollow core conductor concept.

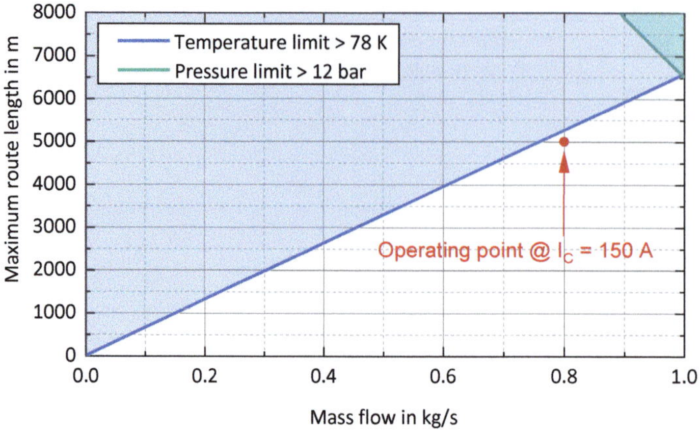

Figure A.15: Representation of the maximum route length in dependency of the mass flow with one-sided cooling for the nominal diameter DN 40 of the hollow conductor Ic = 200 A with omitted conductor layer.

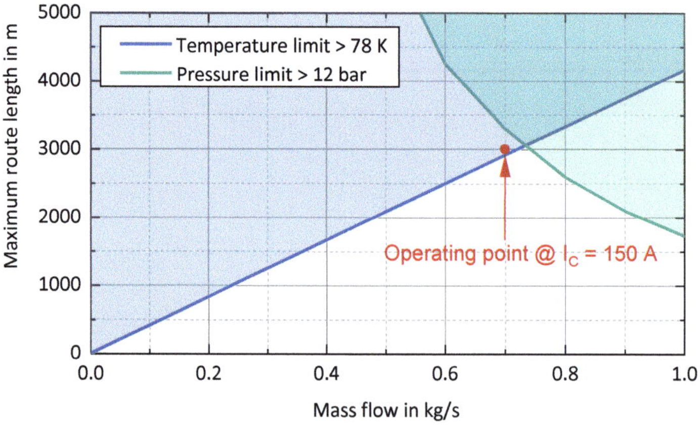

Figure A.16: Representation of the maximum route length in dependency of the mass flow at both sides cooling for the nominal diameter DN 50 of the hollow conductor Ic = 200 A with saved conductor layer.

A.2 Supplements

Table A.1: Maximum route lengths for cooling on one and both sides with reduced cryostat transmission losses depending on the nominal diameter of the inner tube (corrugated tube concept) and the mass flow

Nominal diameter of inner pipe	Mass flow	Maximum route length in m	
		One-sided cooling	Double-sided cooling
DN 32	$\dot{m} = 0{,}5\,\text{kg/s}$	2015 → 3380	3020 → 5070
	$\dot{m} = 1{,}0\,\text{kg/s}$	4030 → 6760	6050 → 10440
DN 40	$\dot{m} = 0{,}5\,\text{kg/s}$	2290 → 3945	3440 → 5920
	$\dot{m} = 1{,}0\,\text{kg/s}$	4585 → 7890	6880 → 11840
DN 50	$\dot{m} = 0{,}5\,\text{kg/s}$	2285 → 4010	3210 → 6020
	$\dot{m} = 1{,}0\,\text{kg/s}$	4575 → 8025	6420 → 12040

Table A.2: Overview of AC losses at line termination, cryostat transmission losses and dielectric losses of total losses as a function of the nominal diameter of the inner tube (corrugated tube concept)

Label	Symbol	Unit	Nominal diameter of inner pipe		
			DN 32	DN 40	DN 50
AC conductor losses at cable end	$P'_{\text{V,L}}(l_{\max})$	W/m	0.67	0.35	0.19
AC shielding losses at cable end	$P'_{\text{V,S}}(l_{\max})$	W/m	0.26	0.25	0.21
Dielectric losses	$P'_{\text{V,0,D}}$	W/m	0.20	0.24	0.30
Cryostat transmission losses	$P'_{\text{V,0,LK}}$	W/m	2.30	2.30	2.30
Sum		W/m	3.43	3.14	3.00

Table A.3: Overview of the superconductor requirement (three-phase) of the conductor layer as a function of the critical current and the inner tube (corrugated tube concept)

Nominal diameter Inner tube	geo. Number of SC tapes Layer 1/layer 2	Superconducting requirement for 1 km route length (three-phase)	geo. Number of SC tapes Layer 1/layer 2	Superconducting requirement for 1 km route length (three-phase)
	I_C = 150 A		I_C = 200 A	
DN 32	26/26	161.5 km	26/26 [a)]	161.5 km [a)]
DN 40	32/33	201.9 km	32/-	99.4 km
DN 50	40/40	248.5 km	40/-	124.2 km

Note: Equation 3.8 results in a minimum electrical number of 32 SC tapes, so that the inner tube variant DN 32 still has to be designed with two HTS layers. The lay angle is assumed to be 15°.

Table A.4: Overview of the superconductor requirement (three-phase) of the shielding layer as a function of the inner tube (corrugated tube concept)

Nominal diameter Inner tube	geo. Number of SC tapes	Superconducting requirement for 1 km route length (three-phase)
DN 32	69	214.3 km
DN 40	73	226.7 km
DN 50	77	239.1 km

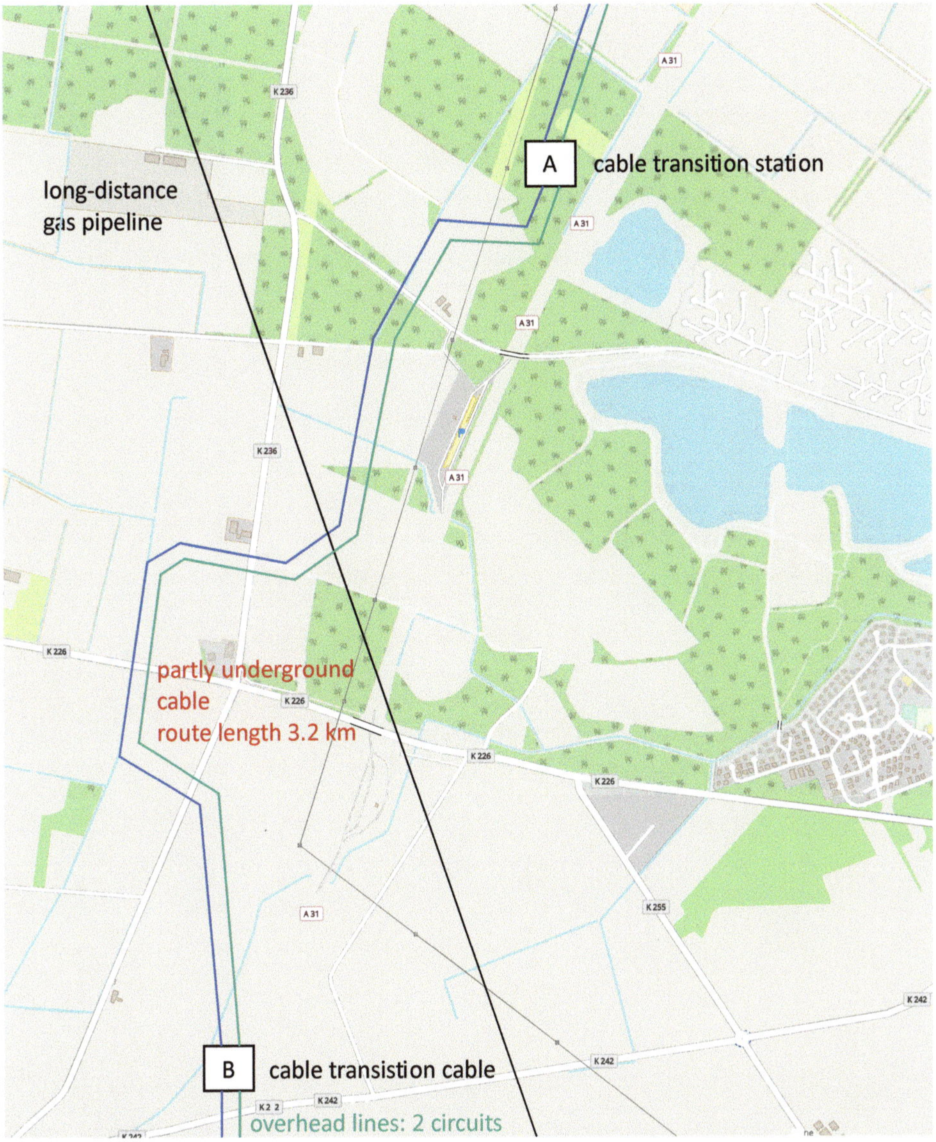

Figure A.17: Overview of the possible route run of the partly underground cable.

Karlsruher Schriftenreihe zur Supraleitung

Karlsruher Institut für Technologie (KIT) | ISSN 1869-1765

Herausgeber: Prof. Dr.-Ing. M. Noe, Prof. Dr. rer. nat. M. Siegel

Die Bände sind unter www.ksp.kit.edu als PDF frei verfügbar oder als Druckausgabe bestellbar.

Band 007 **Christian Barth**
High Temperature Superconductor Cable Concepts
for Fusion Magnets. 2013
ISBN 978-3-7315-0065-0

Band 008 **Axel Stockhausen**
Optimization of Hot-Electron Bolometers for THz Radiation. 2013
ISBN 978-3-7315-0066-7

Band 009 **Petra Thoma**
Ultra-fast $YBa_2Cu_3O_{7-x}$ direct detectors for the THz
frequency range. 2013
ISBN 978-3-7315-0070-4

Band 010 **Dagmar Henrich**
Influence of Material and Geometry on the Performance
of Superconducting Nanowire Single-Photon Detectors. 2013
ISBN 978-3-7315-0092-6

Band 011 **Alexander Scheuring**
Ultrabreitbandige Strahlungseinkopplung in THz-Detektoren. 2013
ISBN 978-3-7315-0102-2

Band 012 **Markus Rösch**
Development of lumped element kinetic inductance detectors
for mm-wave astronomy at the IRAM 30 m telescope. 2013
ISBN 978-3-7315-0110-7

Band 013 **Johannes Maximilian Meckbach**
Superconducting Multilayer Technology for Josephson
Devices. 2013
ISBN 978-3-7315-0122-0

Band 014 **Enrico Rizzo**
Simulations for the optimization of High Temperature
Superconductor current leads for nuclear fusion applications. 2014
ISBN 978-3-7315-0132-9

Band 015 **Philipp Krüger**
Optimisation of hysteretic losses in high-temperature superconducting wires. 2014
ISBN 978-3-7315-0185-5

Band 016 **Matthias Hofherr**
Real-time imaging systems for superconducting nanowire single-photon detector arrays. 2014
ISBN 978-3-7315-0229-6

Band 017 **Oliver Näckel**
Development of an Air Coil Superconducting Fault Current Limiter. 2016
ISBN 978-3-7315-0526-6

Band 018 **Christoph M. Bayer**
Characterization of High Temperature Superconductor Cables for Magnet Toroidal Field Coils of the DEMO Fusion Power Plant. 2017
ISBN 978-3-7315-0605-8

Band 019 **Shengnan Zou**
Magnetization of High Temperature Superconducting Trapped-Field Magnets. 2017
ISBN 978-3-7315-0715-4

Band 020 **Ilya Charaev**
Improving the Spectral Bandwidth of Superconducting Nanowire Single-Photon Detectors (SNSPDs). 2018
ISBN 978-3-7315-0745-1

Band 021 **Juliane Raasch**
Electrical-field sensitive $YBa_2Cu_3O_{7-x}$ detectors for real-time monitoring of picosecond THz pulses. 2019
ISBN 978-3-7315-0786-4

Band 022 **Yingzhen Liu**
Design of a superconducting DC wind generator. 2019
ISBN 978-3-7315-0796-3

Die Bände sind unter www.ksp.kit.edu als PDF frei verfügbar oder als Druckausgabe bestellbar.

Band 023 **Sebastian Hellmann**
Research and Technology Development on Superconducting Current Limiting Transformers. 2019
ISBN 978-3-7315-0804-5

Band 024 **Simon J. Otten**
Characterisation of REBCO Roebel cables. 2019
ISBN 978-3-7315-0904-2

Band 025 **Julia Brandel**
Supraleitende Einzelphotonenzähler: Optimierung der Zeitauflösung und Anwendungsbeispiele aus der Spektroskopie. 2019
ISBN 978-3-7315-0917-2

Band 026 **Dustin Kottonau, Eugen Shabagin, Wescley T. B. de Sousa, Jörn Geisbüsch, Mathias Noe, Hanno Stagge, Simon Fechner, Hannes Woiton, Thomas Küsters**
Bewertung des Einsatzes supraleitender 380-kV-Kabel. 2019
ISBN 978-3-7315-0927-1

Band 027 **Steffen Dörner**
Multifrequenzausleseverfahren von supraleitenden Einzelphotonen-Detektoren. 2019
ISBN 978-3-7315-0961-5

Band 028 **Michael Merker**
Superconducting integrated THz receiver. 2019
ISBN 978-3-7315-0970-7

Band 029 **Wolfgang-Gustav Ekkehart Schmidt**
Superconducting Nanowire Single-Photon Detectors for Quantum Photonic Integrated Circuits on GaAs. 2020
ISBN 978-3-7315-0980-6

Band 030 **Dustin Kottonau, Eugen Shabagin, Wescley de Sousa, Jörn Geisbüsch, Mathias Noe, Hanno Stagge, Simon Fechner, Hannes Woiton, Thomas Küsters**
Evaluation of the Use of Superconducting 380 kV Cable. 2020
ISBN 978-3-7315-1026-0
